Astronomers' Universe

For further volumes:
http://www.springer.com/series/6960

Richard N. Boyd

Stardust, Supernovae and the Molecules of Life

Might We All Be Aliens?

Richard N. Boyd
Ohio State University (Emeritus)
5846 Leona Court
Windsor, CA 95492, USA
richard11boyde@comcast.net

ISSN 1614-659X
ISBN 978-1-4614-1331-8 e-ISBN 978-1-4614-1332-5
DOI 10.1007/978-1-4614-1332-5
Springer New York Dordrecht Heidelberg London

Library of Congress Control Number: 2011942323

Cover illustration: Earth picture from http://earthobservatory.nasa.gov

Printed on acid-free paper

Springer is part of Springer Science+Business Media (www.springer.com)

Preface

In writing this book I gradually realized that I was addressing two very disparate scientific communities: biochemists and astrophysicists, and that these communities are rarely in scientific communication. Furthermore, the astrophysicists were further divided into two additional not-very-interactive communities, nuclear astrophysicists and neutrino astrophysicists. What this forced me to do is try to present what I regard as at least a minimal level of information from each of those communities in the context of the primary subject of this book, the chirality of the amino acids. The biochemists will quickly realize that my allegiances are with the astrophysicists; I hope I have done a reasonably credible job of representing the chemists. Trying to gain a reasonable level of personal credibility in biochemistry has certainly represented a major component of the time I spent in writing this book. In recognition of the wide diversity of topics, and the inevitable nomenclature that goes with each, I've tried to include a pretty complete glossary.

Since this book is about origins, I felt it appropriate to present several types of origins, notably, the Big Bang, the origin of the elements, and finally the origin of the molecules of life in the cosmos and on planet Earth (and quite possibly on other planets). I have tried to present all of these subjects at a level that someone with a minimal background in science, but a strong interest in learning new things, can digest. This might also make the book appropriate for a course for undergraduates who are not science majors: "amino acid chirality for poets" (but with significant additional input from the teacher, probably with the help of the many references to supplement the areas of professorial non-expertise). For those readers who have more of a science background, I hope I've not insulted your intelligence too much; the subjects in this book are sufficiently varied that I can almost guarantee that you will eventually find something discussed in which you're not an expert. In any event, I have tried to describe the subjects of the book in such a way that a class of undergraduates could use the book to truly get a grounding in astrobiology.

Although I have certainly emphasized the model that I, together with my colleagues Toshitaka Kajino and Takashi Onaka, have built, I have also tried to present enough generality in the subjects relevant to the origin of life that students can gain an overall impression of the basic features of the subject. And, of course, I've included some physics background subjects that are favorites of mine, which are relevant to origins, and which can serve as general knowledge for students and other readers who are generally curious about science.

In writing this book I've violated one of the basic tenets associated with attempts to popularize science: don't include equations. I've included lots of equations. In one instance the equation was unavoidable: that was the discussion of the Drake equation, which is pretty basic to discussions of detecting extraterrestrial life. However, my other equations are not the sort of thing you find in algebra books, rather they are more shorthand ways of stating processes or reactions. These are not equations that could be solved for one of the variables; they don't contain variables! I hope the readers will find them useful.

I am deeply indebted to many people for contributing to my knowledge of this subject, and therefore to the content of this book. First and foremost in this regard are my collaborators on the development of the Supernova Neutrino Amino Acid Processing model, Toshitaka Kajino and Takashi Onaka, for their many contributions to our efforts. This book could not have been written without their inputs to our two papers. I must include in my acknowledgements two other people who gave a boost to the early efforts: Isao Tanihata and Reiko Kuroda.

And, finally, I gratefully acknowledge my wife, Sidnee, who although she is a nonscientist, read through the manuscript, and in many instances pointed out the places where the physicist in me had gotten carried away with physics-speak. The places where the physics-speak still lurks are my fault, but the readers can be sure that there are many less such places than there would have been without Sidnee's efforts.

Windsor, CA, USA Richard N. Boyd

About the Author

Dr. Richard N. Boyd was the Science Director of the National Ignition Facility, Lawrence Livermore National Lab from 2007 to 2010 and now serves as a staff physicist at LLNL. He earned his Ph.D. in physics from the University of Minnesota in 1967, and has been a physics professor at the University of Rochester from 1972 to 1978 and a professor of physics and astronomy at Ohio State University from 1984 to 2002. Dr. Boyd also served as a program officer at the National Science Foundation from 2002 to 2006, managing the NSF portfolios in nuclear and particle astrophysics as well as nuclear physics. Following that, he was a visiting professor at the National Astronomical Observatory in Japan.

Dr. Boyd has enjoyed a research career that resulted in more than 200 publications, both experimental and theoretical, and one graduate-level textbook on nuclear astrophysics. He received an Outstanding Scholar award from Ohio State University in 1982, and was named a Fellow of the American Physical Society and of the American Association for the Advancement of Science. He was named an honorary Native American of the Santa Clara Pueblo in 1997, and an Eminent Scientist of the Institute for Physical and Chemical Research in Japan in 1998–1999.

Contents

1. Introduction

Abstract How do you know if something is alive? The answer might seem pretty clear for large creatures, but is not so obvious for single-celled critters. This chapter discusses some of the attempts to define the criteria that determine if something is alive. It also discusses the origin of Earth, the conditions that existed on early Earth, and how those conditions changed so life could eventually survive and flourish. The time-honored Miller–Urey experiment is described; it has been used for decades to describe how some of the molecules of life, specifically the amino acids, might have been produced on early Earth. This chapter then introduces the concept of chirality, or handedness, which is a property of the amino acids, and which raises serious questions about the Miller-Urey based explanation of the origin of life on Earth. Finally, a model is introduced that might explain how the amino acids achieved this chirality, and which allowed them to become the building blocks of more complex molecules of Earthly, and possibly cosmic, life.

1.1 In the Beginning

This book is about origins. Its primary focus is on the origin of life on Earth, or at least on the molecules from which life grew, and perhaps in other places in the Galaxy. But that didn't happen first; the Universe had to get started via the Big Bang, then it had to make the elements that were not made in the Big Bang (and that's nearly all of them!), then the Solar System and Earth had to be created, and then finally, we can consider the origin of life on Earth. And if you hadn't guessed yet, this will be from the perspective of modern science.

The Universe was born 13.7 billion years ago. The age of the Universe is incontestable. It has been determined by a couple of incredibly elegant experiments. The values they get for that age are

R.N. Boyd, *Stardust, Supernovae and the Molecules of Life: Might We All Be Aliens?*, Astronomers' Universe, DOI 10.1007/978-1-4614-1332-5_1,

determined by completely independent, and different, techniques, and they agree extremely well. These experiments are so important that I want to describe some of their details, so they will be discussed in Chap. 2.

The Earth was born 4.55 billion years ago. The elements that exist on Earth, all the carbon, oxygen, silicon, iron, indeed everything except hydrogen and helium, were produced in stars that existed prior to the formation of our Solar System. Because the Universe is so much older than the Earth, those elements are thought to have been produced in several previous generations of stars.

The Earth's age I gave you above is also incontestable; there are at least two ways to determine the Earth's age, and they are consistent with each other. One technique establishes a lower limit on the Earth's age of around 4 billion years by dating the oldest rocks found on Earth. Radiocarbon dating is the most common form of this technique, but we are dealing with ages in the billions of years, and the half-life of ^{14}C, the radioactive isotope of carbon, is much too short (5,730 years) to allow a meaningful measurement of that long an age (the maximum for carbon dating is about 100,000 years). Determining long past ages using radioactive nuclei is a well-established technique; it has been applied to determination of the age of the Universe [1, 2], although newer techniques, to be discussed in Chap. 2, provide a much more accurate answer.

For our present purposes, Nature has provided us with a wonderful crystal: Zirconium Silicate, $ZrSiO_4$. "Zircons" have been found, when they crystallize from magma, to contain uranium (U), but they do not contain lead (Pb) when they crystallize. ^{204}Pb (82 protons, 122 neutrons) is a decay product of ^{238}U (92 protons, 146 neutrons), and ^{205}Pb (82 protons, 123 neutrons) is a decay product of ^{235}U (92 protons, 143 neutrons). Thus the zircons provide a very direct way to measure the Earth's age: measure the abundances of the lead isotopes in rocks that also contain Uranium. The two U isotopes have extremely long half-lives (0.704 billion years for ^{235}U; 4.468 billion years for ^{238}U), and they are known to high accuracy. As the ^{235}U and ^{238}U decay, the amounts of ^{204}Pb and ^{205}Pb in the sample increase. So by measuring the amounts of ^{204}Pb

and ^{205}Pb, and comparing those to the amounts of ^{235}U and ^{238}U, one can determine the age of the rock. The oldest rock appears to have sufficient zircon inclusions to suggest that it is more than 4 billion years old.

How sure are we that this technique is giving correct answers? The results only depend on the half-lives and the accuracy of the abundance measurements. Many tests have been conducted to see to what extent extreme conditions, e.g., the high temperatures encountered in the early Earth, could influence these half-lives. The answer is that such influence is well below the very small uncertainties in the half-lives, and the largest influences (which are still very small) occur in nuclei that decay by a different mechanism than do the U nuclides (the word "nuclides" refers not to just the elements, but to all the isotopes of the elements). Of course, the dating approach can only tell the ages of the rocks; it therefore provides only a lower limit on the age of the Earth. There are other ways to determine the age of the Earth; the number that one achieves with all these techniques is 4.55 billion years [3].

Before leaving the dating story, I should mention that the zircons can also provide an age on life forms that are found in ancient rocks. Unfortunately, the fossils are found in sedimentary rock, and the zircons are definitely not sedimentary rock. However, some sedimentary rock that contained fossils was also found to contain zircons. If the zircons were included in the sedimentary rock at the same time the fossils were entombed therein, then the age would be easy to determine. However, the zircons may well have been added to the sedimentary rock only after they had existed for a while. Moorbath [4] has estimated that the fossils in these rocks are 3.67–3.70 billion years old. I've left out the details, but their ages are very well established.

So what was the Earth like initially? It was a pretty miserable place by any standards. It didn't resemble the current version very much, as it was very hot, had essentially no atmosphere and very little water, and was being continually bombarded by meteorites. Those meteorites are important; they will play a central role in our story. However, in the early history of the Earth, they would have made our planet a pretty unpleasant place to be. In fact, it would have been uninhabitable, as huge meteorites would have created enormous displacements of material, the clouds from which would surely have extinguished any life that might have survived the unbearable temperatures. How do we know this? The Moon was presumably subjected to the same bombardment, and its cratered surface tells us that this was the unfortunate situation for the unfortunate early Earth. This situation has been studied by Sleep et al. [5], among others.

However, this intense meteoritic bombardment ceased about 3.7 billion years ago, again from the evidence we get from the Moon. Although an occasional meteorite still does make it to Earth, their number was greatly reduced from that of the earlier period. Why did this change come about? It appears that Jupiter served as a cosmic janitor. Because its large size, it exerts a huge gravitational pull on objects that come close to it, and that gravitational pull on the objects it encountered as it circled the Sun allowed it to sweep out most of the debris that was originally floating about the inner Solar System. This made it possible for any life that might arise on Earth after the cosmic cleanup was completed to flourish and evolve without too much concern for large scale planetary destruction, although that can still occur from hits from large meteorites that were missed in the Jovian sanitation of the Solar System. Incidentally, it apparently took Jupiter more than a billion years to complete its cleanup of the meteoroids in the inner Solar System.

Of course, the Earth had to cool and it had to develop an atmosphere that contained some oxygen before life, at least life as we know it, could begin. We know Earth also somehow increased its water content to its present level. It is generally believed that water is more or less essential to the existence of life, although that bias could lead us to miss recognizing some forms of life. We will return to this in the last chapter. The importance of liquid water is emphasized in the article by Rothschild and Mancinelli [6]. They note where there is liquid water on Earth, virtually independent of all other conditions, there is life. "Other conditions" here can mean extremes of heat and cold, alkalinity, and a whole host of other possibilities. Of course, the most extreme conditions necessitate the existence of creatures that can withstand the extremes; these are known as "extremophiles."

Plaxco and Gross [7] devote a lot of space in their book to a discussion of the importance of water to life of any kind. I think that discussion is sufficiently important that I'll repeat the most salient points. The human body is about 70% water, so there would seem to be a very strong prejudice in our life forms for that particular liquid. This is seen in virtually every biological process in which our cells and bodies engage; we must have a solvent for all these to happen. Water is as good as it gets.

Furthermore, water has the rather unique property that when it freezes it expands. Thus when lakes freeze, the ice doesn't sink to the bottom, which would make it much easier for such lakes to freeze solid, and would probably eliminate all life forms from such lakes. When water freezes in cells the ice so formed tends to rupture the cells. We will return to this when we discuss extremophiles in Chap. 9.

Water also has an unusual ability to absorb heat. Thus it provides an excellent moderator of climate change. And, of course, it remains liquid over a 100° Centigrade (180° Fahrenheit) temperature range, yet another unusual characteristic, and one that is very accommodating for life. Water also has an unusually large "dielectric constant," which makes it possible to take many chemicals into solution in water. This is an important capability for the functioning of our cells. Earth's water may have arrived in the form of comets, which consist largely of ice, so if enough of them got to Earth, they would have fulfilled that function.

But all of this avoids the question of how life actually began on Earth. There are certainly a number of questions about how that occurred, many of which represent active areas of current research. And there are numerous books that deal with various aspects of that question. I won't repeat much of what the other books cover, but will focus on some issues that may be central to the presence of the amino acids on Earth, which are the building blocks of the proteins on which our lives depend, so in turn are a critical component of the existence of life on Earth. The issues I'll focus on tie together some astrophysical phenomena that occur on macroscopic scales with the microphysics of molecules and even of the constituents of the nuclei of the atoms in those molecules – the very large with the very tiny.

1.2 What Is "Life?"

This might seem like a question that should have an obvious answer, but that does not turn out to be the case. Indeed, scientists have been trying to provide an answer to that question for decades, and their efforts have been preceded by others spanning centuries. A wonderful paper on the definition of life was written by Luisi [8], and I shall use many of his results. Charles Darwin did give some thought to what was required for life to have occurred in the first place from his nineteenth century perspective, and came up with the observation: "The hypothesis of an originary arising of life from the inanimate matter ... can at least offer the advantages to explain natural things by natural pathways, thus avoiding to invoke miracles, which are actually in contradiction with the foundations of science" (quoted from [8]). Subsequently, Alexander Oparin, in the early twentieth century, and culminating in his book "The Origin of Life," [9] attempted to refine the definition by specifying its requisite conditions. He concluded that life was necessarily based on six properties (1) capability of exchange of materials with the surrounding medium; (2) capability of growth; (3) capability of population growth (multiplication); (4) capability of self-reproduction; (5) capability of movement; (6) capability of being excited. He also added some additional properties, such as the existence of a membrane and the "interdependency with the milieu." Although there are many other suggestions of what constitutes life, this should give the reader some idea of the complexity of the attempts to define life [8].

NASA, the US National Aeronautics and Space Administration, obviously has a huge investment of both money and time in searching for life, so it has attempted to provide a definition of its own. Their definition: "Life is a self-sustained chemical system capable of undergoing Darwinian evolution." As noted by Luisi [8], this attempts to move the definition of life into the molecular level. This could evolve even further by relating the definition to ribonucleic acid, or RNA, but now we are getting beyond the scope of this book.

Finally, before I get to criticisms of these definitions of life, let me offer one more: the definition by Paul Davies in his book

"The Fifth Miracle" [10]. I've restructured and condensed his requirements, but basically they include the following:

- Living things are autonomous, that is, they have self-determination. Living things can at least in some sense decide things for themselves. Davies compares a live bird, which can decide where it is going and what it is doing, with a dead bird, which clearly is incapable of making its own decisions. However, the two birds are essentially identical, biologically and chemically, so there is some special feature that the live bird has that the dead one doesn't. This statement doesn't deal with the reasons why one bird is alive and the other is dead, it only recognizes that differences exist.
- Living things reproduce, creating offspring that bear a large resemblance to the parents. This is clearly a requirement for continuation. This also requires that living things have some mechanism for information transfer from one generation to the next.
- Living things metabolize, that is, they process chemicals, thereby bringing energy into their bodies. This requirement exists because living forms use their "food" to produce the energy they need to perform the functions they must in order to continue to live and reproduce.
- Living things have complexity and organization, that is, life forms are composed of many atoms, and the different specialized groups of atoms and molecules must work together in ways that make the systems behave as they must to continue living.
- The complexity that characterizes life must be organized complexity; the separate components must work together. Recall the two birds discussed earlier; although the "systems" in the two birds are essentially identical, they have ceased to work together in the dead bird. In the living bird, the lungs take in oxygen and insert it into the blood stream. The blood nourishes the brain, which serves as director of operations, but it cannot do so without the oxygen that the blood transports to it. So any failure of one of those systems will result in a lack of organization, and death.
- Living things must be capable of developing, or must be able to change in response to demands of the environment. They must be able to evolve, that is, their reproduction must include mechanisms for evolution and natural selection. This is crucial;

it not only determines the ability of any species to change in response to the needs imposed on it by external factors, but it also allows for improvement of any species for survival in its constant competition with other members of its species, and with those of other species. This is done by tiny changes in the DNA of the species, mutations, most of which are destructive, so don't produce survivable offspring, or by imperfections in reproduction. But a few of the changes will improve the ability of the species to survive, and those adaptations will then eventually replace the ones that preceded them. This is how Darwinian evolution capitalizes on the mutations that are produced by, among other things, the cosmic rays that continually rain down on Earth. You and I are in a state of constant mutation!

So how well do these definitions work? Well, they have some obvious problems. For example the NASA definition really defines life for a population, since reproduction often requires two members of most species, so no single entity of these species is alive under that definition. And exceptions are pretty easy to come up with. For example, mules are incapable of reproduction, but are arguably alive (Perhaps stubbornness can replace reproductive capability, although there are certainly examples of politicians who appear to carry both characteristics.). And viruses really complicate the definitions, since they do pretty much fulfill the requirements for categorization as a life form, but are not included as such by everyone. An interesting discussion of this issue is given in the book by Ward [11]. So the bottom line is that it's not so easy to even define what it means to be alive.

Since molecular biologists have gotten involved in the origins of life (how could they avoid it?!), it is natural that modern definitions of life would take us to that realm. Indeed, one edition of the journal Astrobiology contained several papers that dealt with the definition of life. I won't survey all the information in those articles, but will just summarize and reference a sampler of what might be involved in a definition of life based on the microscopic.

Deamer [12] bases his definition on what are generally agreed to be the basic molecules of life: the nucleic acids, that is, the DNA and RNA, and the proteins, all of which are referred to as "biopolymers." He then notes the processes by which the biopolymers are synthesized, that is, by combining amino acids and nucleobases.

His definition of life defines polymer synthesis as the fundamental process that must occur in living entities, using the capabilities of the nucleic acids to store and transmit genetic information, and of the enzymes to act as catalysts for metabolic processes. As with other definitions of life, Deamer includes reproduction as fundamental to his definition, but casts it in molecular terms. Finally, he notes that the cells can evolve.

I've summarized and condensed Deamer's definitions, since if you're not into molecular biology, some of the verbiage can be a bit intimidating. What is most interesting, though, is that many of these properties he lists are not too different from Davies' list of properties, but rather have taken them to a more fundamental, that is, microbiological, level.

Let's talk a bit about basic definitions of life forms, as we will encounter them in subsequent chapters, specifically archaea (ar-KEE-a), bacteria, and eukaryotes (ew-KARR-ee-ohts). Archaea are single-celled microorganisms, and their single cell has no nucleus or many of the other fancy features that human cells, which are eukaryotes, enjoy. However, archaea do have genes and enzymes that catalyze some metabolic pathways, which gives them some similarity, in that regard, to eukaryotes. However, archaea reproduce asexually. Bacteria are also single celled microorganisms and like archaea, their cells have no nucleus. Archaea and bacteria are roughly similar in size and shape, and often grouped together as "prokaryotes," that is, "not eukaryotes." Prokaryotes, simply stated, "are molecules surrounded by a membrane and a cell wall" (http://www.biology.arizona.edu/).

Eukaryotes are considerably more sophisticated than either archaea or bacteria. However, they do have some additional similarities to both. Eukaryotes and archaea are similar in that both have DNA, deoxyribonucleic acid, which governs the genetic machinery in both. However, Eukaryotes also have similarities to bacteria. Eukaryotes differ from both archaea and bacteria, though, in that they have a nuclear envelope [13–15], within which their genetic material is contained. They also have other cellular refinements that archaea and bacteria do not have, notably, many "organelles," organs of the cell, that exist in eukaryotes. All species of large complex organisms, animals, plants, and fungi, are eukaryotes. Although human organisms are eukaryotes, there are many more bacteria than eukaryotes that inhabit the human body! Some of the bacteria are helpful, for example, in aiding in digestion. However, some others can be extremely dangerous; cholera and tuberculosis come to mind.

The first entity that could really be called "living," that is, that would satisfy the criteria that define life, probably had only a few things in common with the modern definition of archaea and of bacteria, and even less with eukaryotes. However, these first cells eventually evolved into the three distinct life-form branches of archaea, bacteria, and eukaryotes. These are further divided into "kingdoms," with archaea and bacteria comprising their own kingdoms, and eukaryotes being divided into four: protista (unicellular protozoans and multicellular algae), plants, fungi, and animals (But note that not everyone agrees on these kingdoms, or even that there are but six of them.).

Cell division is quite different in eukaryotes than in the organisms that do not have nuclei. Although that's a very interesting subject, it's pretty far removed from our story, so we'll leave that for other authors.

Fortunately, for the current book, we don't need to be too fussy about what defines life. As Justice Potter Stewart said in his famous quote about hard-core pornography, "I know it when I see it." Or will we? Some of the single-celled entities discussed in the previous paragraphs might not be so easy to recognize as living entities, especially after having been committed to the fossil record for a couple of billion years. So the answer to the "or will we?" question is not so obvious, and we will get back to it later. What we are presently concerned with is how the complex molecules that we know life requires are selected and propagated in outer space, which indeed we will see that they are. Although the means by which they combine to form life are extremely complex and poorly understood (and this is an incredibly important area of research), we won't have to worry about the details of how that comes about; it clearly does happen!

1.3 The Miller Urey Experiment

Scientists have been trying to answer the question of our chemical origins for at least several decades. Oparin [9], mentioned above, and Haldane [16] have suggested that "a 'primeval soup' of organic molecules could be created in an oxygen-depleted atmosphere (the condition of early Earth) through the action of sunlight. These would combine in ever-more complex fashions until they formed droplets. These droplets would 'grow' by fusion with other droplets, and 'reproduce' through fission into daughter droplets, and so have a primitive metabolism in which those factors that promote 'cell integrity' survive, and those that do not become extinct." (from Wikipedia) Oparin and Haldane concluded that oxygen which developed later on planet Earth would prevent the synthesis of the organic compounds that are necessary for life (although it is clearly essential for the continuation of many Earthly life forms, including humans).

Thus, half a century ago Miller and Urey [17], two chemists at the University of Chicago, performed an experiment to see if they could make amino acids from conditions that might have existed in the early Earth. This was an unusual situation, since the idea for this experiment was Miller's, and he was a graduate student when he began this work. He had to convince his advisor, Urey, that they should do this experiment. However, when the work was published, Urey, already a very well-known chemist, was so concerned that Miller would receive appropriate credit for his work if Urey's name appeared with Miller's on the paper, had Miller publish the paper as the sole author.

There are more than 100 amino acids; 20 of them are the building blocks of the proteins that we require for life. Our bodies make about half of these 20 amino acids, and we get the others that we require from the food we eat. Beyond those 20, the amino acids are not essential for life, and most do not even occur naturally on Earth. Miller and Urey were trying to figure out how the first Earthly amino acids were produced. Their experiment was designed to see if the 20 on which we depend for life might have been created initially in an environment that probably existed early in the Earth's history. What they found was that they could create at least some of the amino acids from a spark discharge in an oxygen poor environment that contained a few basic molecules, consistent with Oparin's and Haldane's suggestion. If one assumes that such an environment could have existed on Earth in its early stages, one can conclude that the amino acids were formed on

Earth in a lightning storm, and that subsequent chemical evolution produced all the molecules that are essential to life.

Is that the end of the story? Not quite. Amino acids also have a peculiar property, called "chirality," that the Miller-Urey experiment does not explain. Chirality is a fancy word for handedness. Your hands have chirality; if you hold them in front of you with the palms facing you it is obvious that they do not look the same; the thumbs point in opposite directions. But if you stand by a mirror so that you are looking at the palm of one hand directly and at the palm of the other in the mirror, you find that now they do look the same. So your hands do have a kind of symmetry—a mirror symmetry. From this you could conclude that there are two forms of chirality: left handed and right handed. It turns out that chirality also exists in molecules, and in fundamental particles, and it's very much like it is with your hands; it's a mirror symmetry. And the two forms have been given the scientific designations of "left handed" and "right handed."

So chirality seems to manifest itself in a variety of situations in nature. Indeed, chirality applies to the amino acids; it has been found that the Earthly amino acids (with one exception, and it is non-chiral) are all left handed! But if you produce them in a spark discharge you produce the same number of left handed and right handed ones, although somehow all the amino acids that exist in us got to be entirely left handed. So there must be some other, or an additional, mechanism by which the amino acids are produced and processed that the Miller-Urey experiment does not address. Therefore, knowing how the amino acid chirality came to be may well provide the key to understanding their origin.

It is now indisputable that amino acids are made in outer space, and that they undergo some chiral selection there; I'll present the scientific evidence for that, which is a result of meteorites that hit the Earth. So we're all extraterrestrial aliens in a sense! Obviously we weren't born on another planet, at least I wasn't, but what I will argue is that the template molecules of the stuff of which we're made were created far from planet Earth, and that the chirality of the amino acids is part of the solution to the question of their origin. It's well established that most of the elements that comprise those molecules were created by stars and expelled into the vast reaches of the cosmos when those stars ended their lives.

But in addition, astronomers have found abundant evidence for complex molecules that have formed in outer space from those elements. And, since some of those are the building blocks of the molecules of life, and one of the mechanisms I will present in this book would establish the chirality that must be created in outer space, there is reason to believe that the basic requirements for life did begin in outer space.

That's what this book is about: it offers an explanation of how one aspect of life came to be the way it is. Obviously that's not as far out as someone's tale about being conceived on another planet and then transported to Earth, but it's a lot more likely to be true. But the scientific origin of the amino acids is also an interesting and exciting story. I hope you'll agree!

1.4 General Background and Definitions

So this book is about the theories that purport to explain how some of the seeds of life, the amino acids, were created in the cosmos, and then transported to Earth to become a critical element in the creation of Earthly life. Many of the components of these theories are not new; scientists as far back as the ancient Greeks have discussed such concepts. Astronomical measurements have provided solid evidence that complex organic molecules (that is, molecules that contain carbon, although that definition is a bit simplistic, as there are inorganic molecules that contain carbon) are continually being created in the cosmos. And, as noted above, a handful of the meteorites that have crashed to the Earth's surface have demonstrated that amino acids and even more complex biologically relevant molecules are created in the cosmos, and can survive their trip to the surface of the Earth.

However, as noted above, Nature has given us a wrinkle that not only suggests that the molecules of life originated in the cosmos, but perhaps even *demands* that this be the case; this evidence is the chirality of the amino acids. We'll talk a lot more about chirality in subsequent chapters. Although the 20 amino acids on which we depend for our lives are all left-handed (with the one exception), there is no obvious reason why at least some of them could not have been right-handed, since experiments performed on Earth that make them have produced equal amounts of each. If we are to understand the origin of life, we must address this important clue as to where it began.

Furthermore the fact that the amino acids are all left-handed doesn't of itself demand that they all have the same handedness, or even that they all have *some* handedness. However, this may well be the case, as Gol'danskii and Kuz'min [18] state in their review article: "It can be concluded ... that the biogenic scenario for the onset of the chiral purity of the biosphere could not, even in principle, have been realized in the course of evolution, *since without chiral purity of the medium the apparatus of self replication could not appear*. This apparatus is a basic process in the self-reproduction of any organism. *Life cannot arise in a racemic* [that is, equally abundant left- and right-handed] *medium*" (Italics are mine.).

So now you know the meaning of one of the terms we will use: "racemic." But we need to introduce another term—"enantiomeric." An enantiomeric medium has more molecules of one chirality than the other, but can have some of each. And, if there is only one chirality, then the medium is "homochiral."

The concept of chirality has existed since the time of Pasteur [19], who noted in 1860 that there was "a 'demarcation line' between life and non-life: the mirror dissymmetry of organisms" (quoted from [18]).

I hope that whets your appetite, not only to find out how the amino acids originated in space, and how they might have been involved in the creation of life, but also to understand more about chirality, and why it is critical to the development of life on our planet. But our story also requires that I tell you how the Universe began, how the elements got produced, why the cosmos favors left-handed molecules. This latter item will require some details of core-collapse supernovae, some information about neutrinos (those ghostly fundamental particles that have no charge and almost no mass), and some basic nuclear physics. I will also explain how the left-handed molecules, once created, get amplified and propagated throughout the Galaxy, and amplified again when they begin their planetary existence. But I'll explain all these things in the appropriate chapters, then assemble all the pieces in the next to last chapter.

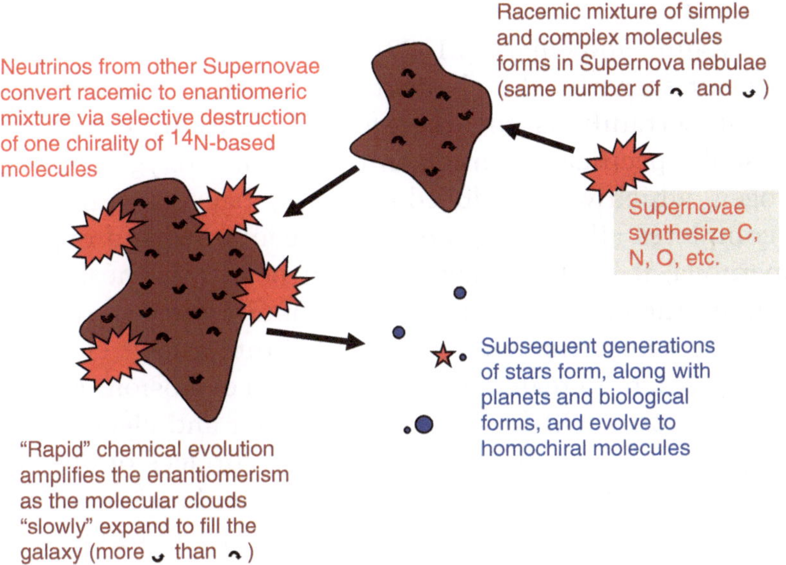
Racemic mixture of simple
and complex molecules
forms in Supernova nebulae
(same number of and)
Neutrinos from other Supernovae
convert racemic to enantiomeric
mixture via selective destruction
of one chirality of 14N-based
molecules
Supernovae
synthesize C,
N, O, etc.
Subsequent generations
of stars form, along with
planets and biological
forms, and evolve to
homochiral molecules
"Rapid" chemical evolution
amplifies the enantiomerism
as the molecular clouds
"slowly" expand to fill the
galaxy (more than)

FIGURE 1.1 Scenario by means of which chiral amino acids are produced. "Racemic" means that there are equal numbers of left- and right-handed molecules. "Enantiomeric" means that there are more of one chirality than the other. "Homochiral" means all the molecules are of the same handedness. The details are explained in the text. (From Boyd et al. [21]. Courtesy of Astrobiology)

If involving the cosmos in determining something that seems so Earthbound seems improbable to you, let me quote from a review article by Bonner [20] (which by the way is an excellent review article): "The logical conclusion is that the source of terrestrial chirality must then have been extraterrestrial, and furthermore that it must have been capable of providing an ongoing influx of chiral molecules having uniform chirality. Only in this way, and not through the gradual accumulation of chiral molecules after statistical fluctuation in the prebiotic soup could the 'chiral catastrophe' of global symmetry breaking occur during Earth's turbulent prebiotic era."

Figure 1.1 gives you some indication of where we are going in general, and what my favorite theory for explanation of amino acid chirality looks like. Any theory of amino acid creation that has them being produced in the cosmos (and I will discuss others) will involve many of these same stages, so this theory provides a useful discussion of the features of these theories that purport to explain amino acid chirality that applies to more theories than just ours. This particular theory is from work I did with my colleagues Kajino and Onaka [21, 22].

In our scenario, the elements are mostly produced in stars, and that certainly is the case for the ones in which we are interested, specifically carbon, nitrogen, and oxygen (except for hydrogen, which was produced in the big bang). When a massive star completes all its stages of stellar evolution and explodes in a supernova, it seeds the cosmos with newly synthesized nuclei (which are the centers of atoms, and contain most of the mass of the atoms). These elements are expelled into a huge cloud, called a nebula, which is created by the supernova explosion. As the nebula cools, atoms will form from the nuclei and electrons in the nebula, then molecules will form on the surfaces of dust grains, or on meteoroids ("meteoroids" are the chunks of rock and possibly ice that exist as part of stuff of outer space; when they pass through the Earth's atmosphere and hit the Earth's surface they are designated "meteorites"). Sometime later, a second star completes all its stages of stellar evolution and becomes a supernova, processing the molecules produced from the detritus of the first star through mechanisms we will discuss later.

But I'll show that it may well be that not any old massive star will do, but that a particular type of star, an especially massive one, that has become a "Wolf-Rayet" star prior to its explosion as a supernova, is required. These stars will ultimately end up as black holes, a property that the Boyd–Kajino–Onaka model [21, 22] requires. When the Wolf-Rayet star explodes it usually produces an intense magnetic field and always produces a huge neutrino flux (we'll have a lot more to say about neutrinos). Some of the dust grains and meteoroids that were floating around the Galaxy will be close enough to the supernova that they, and the molecules they contain, will be affected by the intense neutrino flux and the magnetic field of the supernova. This combination, in interacting with the molecules, produces a selective destruction of one orientation of the ^{14}N nuclei (an isotope of nitrogen, the nucleus of which is comprised of seven protons and seven neutrons) that exist in the molecules that formed; this selects a particular chirality of those molecules. This chirality selection is then amplified by chemical replication that takes place on the molecules that have collected on dust grains that exist in the nebula and its surrounding region.

Then, mixing of the constituents of the Galaxy produces the same chirality selection throughout much of the Galaxy. Next, some of the enantiomeric molecule-containing meteoroids will fall onto Earth and other planets, seeding them with molecules of that chirality. Finally more amplification drives all of those molecules to homochirality, the condition in which only one chirality exists. This is what we have found to be the case on planet Earth.

So the idea that the chirality of the amino acids arose in outer space, beginning with production of the elements, followed by formation of the molecules on dust grain surfaces, and then molecular amplification, isn't new. What is new is the mechanism by which the molecules got processed into the chiral state that has been found to exist.

But this isn't the only model that has been invoked to explain amino acid enantiomerism. Several excellent books have written that attempt to describe how life began. These include two books by Davies [10, 23] and one by Plaxco and Gross [7]. However, these books pay very little attention to the issue of amino acid chirality, which I believe to be central to understanding how and where the molecules of life were created. A book by Meierhenrich [24] and the review articles by Rode et al. [25], and by Avalos et al. [26] give much more attention to the origin of chirality; we will return to some of the work of both the Meierhenrich and Rode groups later on. Another book, by Guijarro and Yus [27], deals with molecular chirality as its primary subject. And, as mentioned previously, articles by Bonner [19, 28] deal directly with the issues associated with amino acid chirality.

To find out for yourself how the molecules of life got started, read on. I'll try to give you a sampler, and critiques, of some of the theories that purport to explain how life as we know it, or in whatever form it might have taken, began.

References

1. J.J. Cowan, F.-K Thielemann, and J.W. Truran, Radioactive Dating of the Elements, Ann. Rev. Astron. Astrophys. 29, 447 (1991)
2. J.J. Cowan, A. McWilliam, C. Sneden, and D.L. Burris, The Thorium Chronometer in CS 22892-052: Estimates of the Age of the Galaxy, Astrophys. J. 480, 246 (1997)

3. C. Stassen, The Age of the Earth, The TalkOrigins Archive, http://www.talkorigins.org/faqs/faq-age-of-earth.html (2005)
4. S. Moorbath, Palaeobiology: Dating Earliest Life. Nature 434, 155 (2005)
5. N.H. Sleep, K.J. Zahnle, J.F. Kasting, and H.J. Morowitz, Annihilation of Ecosystems by Large Asteroid Impacts on the Early Earth, Nature 342, 139 (1989)
6. L.J. Rothschild and R.L. Mancinelli, Life in Extreme Environments. Nature 409, 1092 (2001)
7. K.W. Plaxco and M. Gross, *Astrobiology*, Johns Hopkins University Press, Baltimore, 2006
8. P.L. Luisi, About Various Definitions of Life, Orig. Life Evol. Biospheres 28, 613 (1998). Use of quote courtesy of Springer Publishing
9. A.I. Oparin, The Origin of Life on Earth, Oliver and Boyd, London (1957), originally 1936 in Russian, translated by Ann Synge
10. P. Davies, *The 5th Miracle: The Search for the Origin and the Meaning of Life*, Simon and Schuster, NY, 1999
11. P. Ward, *Life as we Do Not Know It*, Penguin Books, NY, 2005
12. D. Deamer, Astrobiology—Can Life Be Defined?, Astrobiology 10, 1001 (2010)
13. R.M. Youngson, *Collins Dictionary of Human Biology*, Glasgow, Harper-Collins, 2006
14. D.L. Nelson and M.M. Cox, *Lehninger Principles of Biochemistry* (4th edition) New York, W.H. Freeman (2005)
15. E.A. Martin, *Macmillan Dictionary of Life Sciences* (2nd edition), London, Macmillan Press (1983)
16. J.B.S. Haldane, Pasteur and Cosmic Symmetry, Nature 185, 87 (1960)
17. S.L. Miller, The Production of Amino Acids Under Possible Primitive Earth Conditions, Science 117, 528 (1953); S.L. Miller and H.C. Urey, Science 130, 245 (1959)
18. V.I. Gol'danskii and V.V. Kuz'min, Spontaneous Breaking of Mirror Symmetry in Nature and the Origin of Life, Sov. Phys. Usp. 32, 1 (1989). Use of quotes courtesy of American Institute of Physics: DOI:10.1070/PU1989v032n01ABEH002674
19. L. Pasteur, Researches on the Molecular Asymmetry of Natural Organic Products, Alembic Club Reprints No. 14, p. 29, Edinburgh: E.&S. Livingstone Ltd. 1948
20. W. Bonner, The Origin and Amplification of Biomolecular Chirality, Orig. Life Evol. Biosphere 21, 59 (1991). Use of quote courtesy of Springer Publishing
21. R.N. Boyd, T. Kajino, and T. Onaka, Supernovae and the Chirality of the Amino Acids, Astrobiology 10, 651 (2010)

22. R.N. Boyd, T. Kajino, and T. Onaka, Stardust, Neutrinos, and the Chirality of the Amino Acids, Int. J. Mod. Sci. 12, 3432 (2011)
23. P. Davies, *The Eerie Silence: Renewing Our Search for Alien Intelligence*, Houghton, Miflin, Harcourt, NY, 2010
24. U.J. Meierhenrich, *Amino Acids in Chemistry, Life Sciences, and Biotechnology*, Springer, Heidelberg, 2008
25. B.M. Rode, D. Fitz, and T. Jakschitz, The First Steps of Chemical Evolution Towards the Origin of Life, Chemistry and Biodiversity 4, 2674 (2007)
26. M. Avalos, R. Babiano, P. Cintas, J.L. Jinenez, J.C. Palacios, and L.D. Barron, Absolute Asymmetric Synthesis Under Physical Fields: Facts and Fictions, Chem. Rev. 98, 2391 (1998)
27. A. Guijarro and M. Yus, The Origin of Chirality in the Molecules of Life, RSC Publishing, Cambridge (2009)
28. W. Bonner, Parity Violation and the Evolution of Biomolecular Homochirality, Chirality 12, 114 (2000)

2. What is the Origin of the Lightest Elements?

Abstract Although the birth event of our Universe occurred 13.7 billion years ago, it left enough signatures about its details that scientists are quite confident in our understanding of the basic features of that event. The first hints of the Big Bang came from astronomers, as discussed in this chapter. More recently, two incredible experiments, the Supernova Cosmology Project and the Wilkinson Microwave Anisotropy Probe, have determined the parameters that govern our Universe in exquisite detail. One longstanding paradox is also discussed, and shown to be solved by the Big Bang model. Finally, we explore the nuclear reactions that made a few light nuclei in the few minutes that followed the Big Bang. The abundances for these nuclei obtained by observational astronomers are compared to the calculations of the nucleosynthesis that occurred just after the Big Bang.

2.1 The Big Bang

We need to begin our story with the basic constituents of molecules, that is, atoms. So where do the atoms come from? Most of the elements are made in stars. But things are not quite that simple. The elements hydrogen and helium comprise 99% of the mass of the universe that isn't made of exotic stuff (that is, dark matter or dark energy, if you're a physics aficionado), and they were mostly produced in the Big Bang. And the birth event of our Universe is certainly the origin of everything we know, so let's begin our story with a discussion of the Big Bang. That name originated with Fred Hoyle, who actually believed in a "steady state universe," that is, one that didn't have a birth event. Hoyle intended the name as a

R.N. Boyd, *Stardust, Supernovae and the Molecules of Life: Might We All Be Aliens?*, Astronomers' Universe, DOI 10.1007/978-1-4614-1332-5_2,

pejorative comment on the model of his competition. Of course, we now know that the name caught on, and the birth event of our Universe is now a well-documented scientific paradigm. Our primary interest will be in Big Bang Nucleosynthesis, BBN, but before we describe that, let's back up a few minutes to the events that preceded BBN.

Thirteen billion seven hundred million years ago an extraordinary event occurred: our Universe was born. This is well documented by many observations, but the Supernova Cosmology Project, SCP, and the Wilkinson Microwave Anisotropy Probe, WMAP, stand out as the modern incarnations of these efforts. The SCP was headed by Saul Perlmutter (who won the 2011 Nobel Prize in physics for his efforts), and WMAP by Charles Bennett.

However, a very prominent forerunner of these occurred in the early twentieth century as a result of astronomical observations by Vesto Slipher and their interpretation by Edwin Hubble. Hubble was an interesting character, noted in his early life more for his athletic prowess than his academic abilities. He once won seven first places in a track meet, and he dabbled in amateur boxing for a time. He also got a law degree before serving in the military, and then getting his Ph.D. Getting back to astronomy, Slipher had noted that the light from some galaxies appeared to be "red shifted," that is, the characteristic wavelengths of the light from those galaxies could be identified as originating from emissions of photons—those particles of light—from atoms of hydrogen, but they were shifted toward longer wavelengths. Since hydrogen is the most abundant element in the Universe, it is appropriate to show some of the characteristic wavelengths that are emitted by hydrogen atoms; no other element emits light at those same wavelengths. These are seen in Figure 2.1.

Figure 2.1 shows the different series (or groups) of emissions of photons when electrons change from one allowed state to another, that is, these are the result of transitions between specific energy levels. The Lyman series results from transitions to the lowest lying energy level, called the ground state. The Balmer series is from transitions to the next highest energy level, called the first excited state. The Paschen series is to the second excited state, and continues beyond the scale to the right. Each series has many more lines, but they pile up at the left most line in each

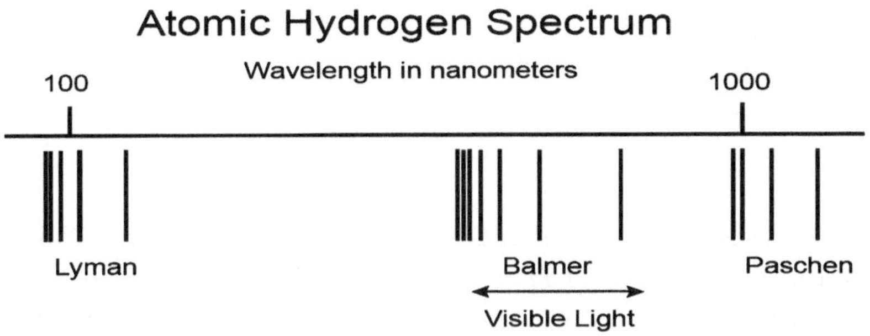

FIGURE 2.1 Characteristic emissions of hydrogen atoms. The electrons in atoms can only exist at well-defined energies which result from well-defined quantum mechanical "states," which are different for the different elements. Transitions between those states produce the characteristic emissions.

series. Those indicated in Figure 2.1 as being in the visible light region would be observed in the laboratory, that is, these are not red shifted. The Lyman series is shifted into the visible part of the spectrum in highly red shifted objects; these are the emission lines that were observed by Slipher.

We will give a more thorough discussion of wavelength in Chap. 4. For now, we will just observe that "light" is electromagnetic radiation, and that it is characterized by an oscillating electric field and an oscillating magnetic field. The oscillations occur in both space and time. The wavelength is the distance over which a wave repeats itself. Visible light has a wavelength of around 5×10^{-7} m, or 1/2 of one millionth of a meter. The above-mentioned characteristic wavelengths of the light from distant galaxies had to have been a result of a Doppler shift, that is, the fact that the galaxies were moving away from us. This is something that everyone experiences in hearing a train whistle or a police siren: the frequency of the sound is higher (and that means that the wavelength is shorter) when the train or police car is moving toward us, then it drops as it passes us. The same effect applies to light. This ultimately led Hubble to conclude that all galaxies were moving away from all other galaxies in the Universe, that is, that the Universe was expanding. He also concluded from the amount of the red shifts that the more distant the galaxies were, the faster they were receding. This led to "Hubble's law:"

$$v = HR,$$

where v is the velocity of recession between galaxies, R is their separation distance, and H is the constant of proportionality, the Hubble constant. This law is a very simple looking equation, but it has profound consequences: it says that the farther an object is from us the more rapidly it will be receding from us, and this applies to every pair of objects in the Universe! This law prevailed for more than half a century, albeit with a large uncertainty on the value of the Hubble constant.

It's not so easy to envision what this looks like in three dimensions, but if you can for the moment think of our Universe as being just two dimensional, then you can think of it as existing only on the surface of a balloon. If you mark galaxies on the balloon, then blow it up, you will see that every galaxy is receding from every other galaxy. However, I don't know many people who can accurately conceive of this in three dimensions, so if you're having trouble, you're not alone.

Indeed, determination of the Hubble constant led to a major irony of twentieth century science. Two major groups had been performing observations and analyses to determine H. One group, headed by the French astronomer Gerard deVaucoleurs, consistently obtained values around 100 km/s/megaparsec (a parsec is an astronomical unit of distance, and is 3.6 light years, or 3.1×10^{13} [31 trillion] kilometers). The other group, headed by Alan Sandage, an American astronomer, consistently obtained values of around 50 km per second per megaparsec, and the uncertainties on their respective values were much smaller than the differences between them.

In science, when you have two results as discrepant as these, the last thing you would do is average them, since one of them is surely incorrect. Of course, both could be incorrect, and that turned out to be the case here. The modern value for H is 70.5 km per second per megaparsec (WMAP website), close to the average of the Sandage and deVaucoleurs results, and certainly the average of the two results within their uncertainties.

2.2 The Supernova Cosmology Project

However, the SCP found, via very detailed measurements done in the 1990s, that the Hubble constant was not constant! Hubble's law had become so ingrained in astronomy that the red shift of distant objects was used to infer their distance. So if one were to check Hubble's law one would need some independent distance indicator. If you think about making an astronomical observation you will quickly realize that it is easy to locate objects on an up–down and left–right plane, but determining the distance to an object, the third dimension, is trickier. What one needs is a class of objects that always exhibit the same intrinsic brightness, or which produce some other observable quantity that allows determination of the intrinsic brightness. Then the observed brightness allows one to infer the distance to the object, since the observed brightness falls off as the inverse square of the distance to the object. One class of objects in the latter category is Cepheid variables, stars for which their brightness oscillates with a frequency that can be related directly to the intrinsic brightness. So astronomers can measure the frequency of oscillation of a Cepheid, and thus determine its intrinsic brightness. And comparing that to the observed brightness then gives the distance to the star. Unfortunately, Cepheids are not especially bright stars, so some other "standard candle" needed to be found for making measurements at the huge distances that characterize cosmology.

Such objects are Type Ia supernovae. These are extremely bright exploding stars that are all essentially the same mass before they explode, and therefore, since they explode by thermonuclear runaway and blow up the entire star, and the nuclear processes are essentially the same for all Type Ia supernovae, they have nearly identical intrinsic brightness. The SCP utilized Type Ia supernovae for its standard candles.

The simplest description of the Universe would be that it formed in a giant explosion of the entire Universe, and has been expanding, and slowing its rate of expansion, ever since. The reduced rate of expansion would result from the gravitational attraction of all the constituents of the Universe on each other. What the SCP found, however, was that although the Universe is expanding, the expansion was speeding up, not slowing down. This suggests that there is something that acts like a "negative gravity," that is, that it does exactly the opposite of what gravity does. This has been dubbed "dark energy." Its existence actually harks back to Einstein, who included it in his general equations and called it a cosmological constant. He later referred to it as his greatest mistake!

So nearly a century later we have come to realize that, as was often the case, Einstein was way ahead of his time, and way ahead of his fellow scientists. In any event, determining what the dark energy is and understanding why it acts as it does will constitute one of the primary objectives of scientists for at least the next decade.

The results from the SCP are summarized in Figure 2.2, which plots the "effective m_B," (which is just the observed brightness of

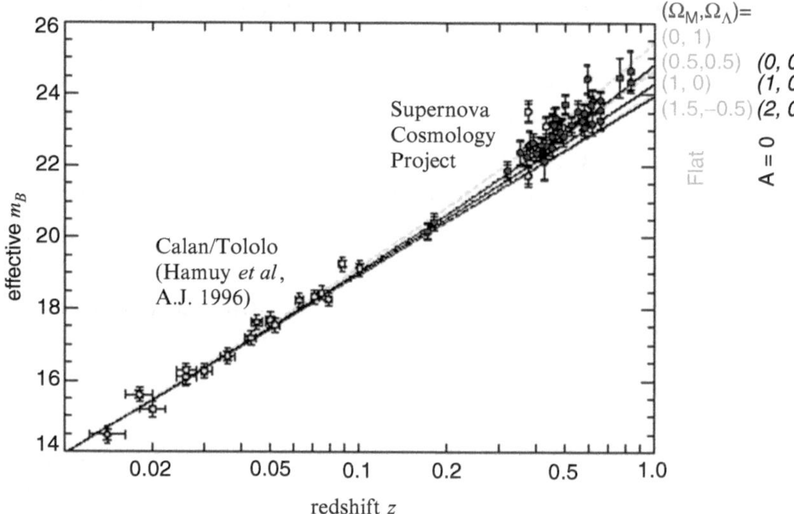

FIGURE 2.2 Hubble diagram, showing effective m_B, the effective peak brightness of the supernovae, versus redshift z, for 42 high-redshift type Ia supernovae from the Supernovae Cosmology Project, and 18 lower-redshift type Ia supernovae from the Calan/Tololo Supernova Survey [1]. Several outliers that were not included in determining the fits to the data are indicated as open circles (although their inclusion did not affect the conclusions). The solid curves are the theoretical effective $m_B(z)$ for a range of cosmological models with zero cosmological constant Ω_Λ and varying energy density of the Universe Ω_M, $(\Omega_M,\Omega_\Lambda)=(0,0)$ on top, (1,0) in the middle, and (2,0) on the bottom. $\Omega_M = 1$ is the critical density of the Universe, as defined in the text. The dashed curves are for a range of cosmological models: $(\Omega_M,\Omega_\Lambda)=(0,1)$ on top, (0.5, 0.5) second from top, (1,0) third from top, and (1.5, –0.5) on the bottom. The high-redshift data are clearly seen to favor a nonzero Ω_Λ. (Reprinted from Boyd [2]. Originally from Perlmutter et al. [3]. Courtesy of IOP Publishing, and of Saul Perlmutter)

each star) on the y-axis versus "redshift" on the x-axis. Redshift is simply related to the "time from Today," or in astronomers' jargon, "lookback time." The scale of the Universe characterizes the expansion of the Universe. The several curves represent the Universal expansion in terms of the different assumptions about the different cosmological parameters that characterize the Universal expansion.

Figure 2.2 shows that the data lie up and to the left of the line labeled (1,0), which is where the data would be if the energy density of the Universe were equal to the "critical value" and the cosmological constant was zero. By critical value we mean that the Universe would continue to expand forever, but slow down at a rate such that it would only stop expanding at infinite time. Thus the data do not favor a scenario in which the Universe is expanding at a decreasing rate resulting only from mutual gravitational attraction. Rather the Universe appears to favor a scenario in which the Universe is expanding, and the expansion *is accelerating*. So the thing to take away from this is that the SCP data do not support the model that had prevailed for more than half a century, but do support a considerably more complicated universe—one that contains stuff that scientists, excepting Einstein (with his cosmological constant), had not imagined previous to the SCP results.

2.3 The Wilkinson Microwave Anisotropy Probe

Although the SCP is probably the most direct way to check the veracity of Hubble's law, it is not the only experiment done to determine cosmological parameters. As mentioned above, the Wilkinson Microwave Anisotropy Probe, WMAP, was the prime example of another class of projects to study cosmology; it was directed toward the 2.7 K (this corresponds to –270.5° Centigrade, 4.9° Rankine, and –454.8° Fahrenheit. This is pretty cold no matter what temperature scale you use!) cosmic microwave background radiation. This is the electromagnetic radiation—the photons—left over from the Big Bang. It was first discovered by accident by Arno Penzias and Robert Wilson, two scientists at Bell Laboratories in New Jersey, as they were trying to develop an extremely sensitive antenna.

Penzias and Wilson were unable to eliminate some background noise, despite heroic efforts to do so. Fortunately, down the road a few miles at Princeton University was Robert Dicke, a theoretical cosmologist, who explained to Penzias and Wilson that they had discovered the relics of the radiation produced in the Big Bang. This radiation was very hot at the time of the Big Bang, but as the Universe expanded, the wavelength of the radiation lengthened with the expansion of the Universe. And larger wavelengths mean less energetic radiation. So this radiation is now extremely cold. It should also be noted that George Gamow had predicted that such a background should exist many years earlier. Penzias and Wilson won the Nobel Prize for their discovery. A much more elegant experiment was performed in the late twentieth century by a team lead by Smoot and Mather [4]; they measured the 2.7 K cosmic microwave background radiation in detail. Smoot and Mather were also awarded the Nobel Prize for their efforts.

However, the Smoot–Mather measurement has undergone an incredibly sophisticated improvement with the WMAP, which was designed to measure temperature fluctuations in the background radiation rather than the temperature itself. It turns out that the density in the early Universe was blotchy, and the sizes and densities of the blotches tell us a great deal about the status of the Universe at the time that the electrons were captured on nuclei to form neutral atoms. Prior to this, the electrons had been free because the temperature of the Universe had been too high, and therefore the density of photons with sufficient energy to ionize the atoms was too great, to permit atoms to exist. The measurement of the blotches can provide tests of theories of how the early Universe formed and evolved. WMAP was able to measure the fluctuations to a few parts in a million; an incredible achievement.

A standard mathematical technique was then used to generate the curves that represent the oscillations in the data shown in Figure 2.3. The x-axis is the size scale of the fluctuations on the cosmic microwave background, as indicated, and the y-axis is essentially the magnitude of the fluctuations in the cosmic microwave background radiation at that angular scale. These representations show that the baryon density of the Universe—baryons are the protons and neutrons that comprise the nuclei of the atoms of which we are made—constitute less than 5% of the mass-energy

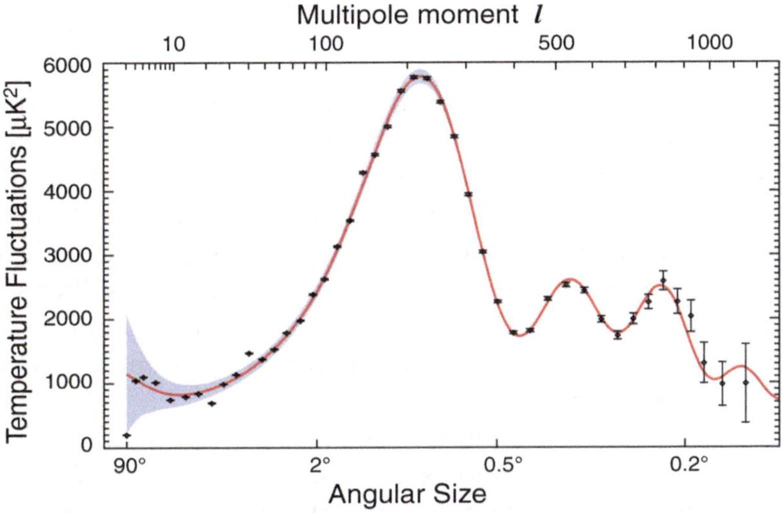

FIGURE 2.3 The *WMAP* angular power spectrum for 7 years of data. The *WMAP* temperature fluctuations are shown in microKelvin (millionths of a Kelvin degree) as a function of the angular size of the fluctuations The best fit Cold Dark Matter with Cosmological Constant model is shown. Author: The WMAP/NASA Science Team. Sponsor: National Aeronautics and Space Administration. (Courtesy of Charles Bennett)

of the Universe. This should produce some level of humility; not only are we not the center of the Universe or even of the Galaxy, we're less than 5% of the stuff from which the Universe is made! This is very accurately determined; the second peak from the left shown in Figure 2.3 is very sensitive to that value.

The WMAP data also showed that 23% of the Universe is dark matter; this is stuff that interacts very weakly with most probes that one might devise to look for it, but does produce a gravitational pull on galaxies, so is obviously present from the motions of galactic constituents. The third peak is especially sensitive to the amount of dark matter. Finally, the dominant component of the Universe's mass-energy, 72% of it, is the dark energy, as determined from both the SCP and the WMAP results. The two experiments also determined that the age of the Universe, to high accuracy, is 13.7 billion years, and the Hubble constant is 70.8 km per second per megaparsec. The original WMAP publication was by Bennett et al. [5], but the most recent results from WMAP at the time this is being written can be found in Jarosik et al. [6].

2.4 Olber's Paradox

There is an interesting argument that shows that the steady state Universe, Fred Hoyle's favorite cosmological theory, can't be correct, or at least has serious problems. This is known as Olber's Paradox. Simply stated, this asks why the night sky isn't bright instead of dark with speckles of starry light? Well, maybe it never occurred to you that the night sky might be bright! Olber's Paradox was promoted in the nineteenth century by astronomer Heinrich Olbers, although the idea behind it was apparently realized as early as the sixteenth century by Kepler.

So why might the night sky be bright? Consider the left-hand side of Figure 2.4, which shows a star seen by observer "O," who is distance d away from it. Suppose now that the same star is distance 2d from the observer; then, since the intensity of the light from the star falls off as $(1/\text{distance})^2$, the star will appear to be 1/4 as bright.

Now consider the right-hand side of Figure 2.4, in which are shown two thin shells of space, each of thickness Δ, one a distance d from the observer and the other a distance 2d from the observer. The observer's telescope will be able to see an angular opening of Θ, which will produce an image of the two disks, the nearer one of diameter Θd, and the more distant one of diameter 2Θd. Thus the nearer disk will have an area proportional to d^2, and the more distant one will have an area proportional to $(2d)^2$ or $4d^2$. So

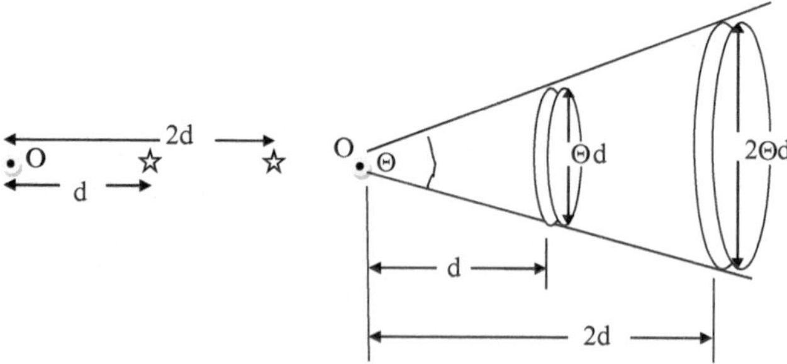

FIGURE 2.4 *Left* Observer "O" in relationship to two stars of the same intrinsic brightness, but the more distant one at twice the distance as the nearer one. *Right* Observer looking through a telescope at two disks that are subtended by the same angle Θ. The disk that is d away has a diameter of Θd, and the one that is 2d away has a diameter of 2Θd.

the volume of the more distant disk, if we now include the third dimension, the Δ, will be four times that of the nearer disk and, assuming that the density of stars is constant, will have four times as many stars as the nearer disk. Each of those stars will appear to have 1/4 of the intensity of the stars in the nearer disk, but there are four times as many of them, so the light from the more distant disk will be exactly equal to that of the nearer disk. If you keep adding disks out to infinity, you should just keep adding to the light seen by the observer, and you will make things bright indeed! Note that I assumed that the Universe was both uniformly populated with stars and that it is infinite. These are assumptions that accompany the Steady State Universe theory.

Okay, so the night sky isn't bright, which means there must be something wrong with the assumptions that went into Olber's Paradox. First, we know that the Universe is not infinite, although it is pretty huge, so that may not resolve the problem. Secondly, the Universe is not static, that is, it is not in a steady state. In fact, we know that it is expanding, and this will increase the wavelengths of the radiation from the distant stars, which also decreases the energy of the photons. In fact, the energies of the light from sufficiently distant stars will be so low as to be irrelevant. Furthermore, the Universe is, in a sense, young, in that the light from distant stars hasn't had time to reach us. And this certainly would not be the case if we had a steady state Universe [6, 7]. So Olber's Paradox really isn't a paradox at all when viewed from the perspective of modern cosmology.

So we know what the baryonic matter is, but we don't know what comprises the dark matter, and we don't have any idea what the dark energy is!

2.5 Big Bang Nucleosynthesis (BBN)

Now, back to our effort to describe what nuclides are synthesized in the Big Bang. What I will do is take you through the nuclear reactions that make the nuclei which were produced in the Big Bang. Along the way we will encounter some physics conservation laws, but I'll explain those also as we proceed. Actually very little was synthesized, due to a couple of nuclear quirks: there are no stable mass 5 or mass 8 nuclides. Mass 5 would be ^{5}He or ^{5}Li, and mass 8 would be ^{8}Be. ^{8}Be is close to being stable, but it is very short lived; it lives 10^{-16} (one ten millionth of one billionth) seconds. The two mass 5 nuclei are much too unstable to live for even a short time. These facts make it virtually impossible to form anything in BBN except ^{2}H, ^{3}He, ^{4}He, and ^{7}Li, and the ^{7}Li is so difficult to produce that its abundance is extremely small. I'll explain what the various nuclides are comprised of below.

But let's see how these nuclides are made. Seconds after the Big Bang, the only nuclear particles were protons and neutrons. As the Universe was expanding, it was also cooling, but it needed to cool quite a bit before the very first reaction could take place. That reaction is

$$^{1}\mathrm{H} + \mathrm{n} \rightarrow {}^{2}\mathrm{H} + \gamma,$$

where ^{1}H and n refer to the protons and neutrons, ^{2}H is a heavy hydrogen nucleus—a deuteron, comprised of a proton and a neutron, and γ is a gamma-ray, a particle of electromagnetic energy. A gamma-ray is a very energetic form of electromagnetic radiation, that is, a particle of ordinary light; all such particles are called photons. A photon is a necessary component of that reaction in order for it to conserve energy. Energy conservation is a major law in physics. This includes not only the energy of motion, but of mass energy, that is, $E = mc^2$, Einstein's famous equation. So energy conservation says that the sum of the mass energies and the energies of motion of the particles on the left-hand side of the equation must equal those same quantities on the right-hand side. However, the Universe had to cool enough for the photons in the above reaction to lose enough energy that they would not run that reaction backwards, that is, so that they would not destroy ^{2}H as rapidly as it was made.

There are two more conservation laws that we need to attend to in this and all the other reactions that follow. The first is "charge conservation." In the above equation, the proton has a charge of +1 and the neutron has zero charge. On the right hand side, the deuteron also has a charge of +1, and the gamma-ray has zero charge. So each side has a charge of +1, and charge is conserved. If there had been an electron in the equation, its charge would have counted as −1. The other law that must be satisfied is "baryon conservation." For our purposes, baryons are just protons and neutrons and nothing else (there are others, but they occur at higher energies than we will be dealing with, so they won't concern us), so the number of protons and neutrons on the left side of the equation, including those that exist in nuclei that contain both protons and neutrons, has to be equal to that number on the right side. (By the way, baryon conservation is thought by particle physicists to be violated, but only at an incredibly tiny level. For our purposes, baryon conservation applies.) Gamma rays are not baryons, so there isn't a conservation law that affects them, aside from conservation of energy. The deuteron is an unusually loosely bound nucleus. Generally it takes around 8 MeV—million electron volts—a unit of energy that is appropriate to nuclei, to liberate a single proton or neutron from a nucleus, but the deuteron can be broken into its constituent proton and neutron with only 2.2 MeV, a pretty small amount of energy by nuclear standards.

So, the deuteron really was the bottleneck that required the Universe to cool before BBN could begin. However, once ^{2}H began to be formed, BBN began in earnest. There was also a bit of a contest going on. A free neutron is not a stable particle; it decays to a proton, an electron, and a neutrino (technically, an electron antineutrino; we will get to neutrinos later on) with a half-life of just a little over 10 min. So after 10 min you will have only half the neutrons that you had before that period started. However, most of the neutrons will get captured into nuclei, and they are stable in their nuclear homes, provided that the resulting nucleus is stable.

The reactions that convert protons and neutrons into ^{4}He nuclei—comprised of two protons and two neutrons—are as follows:

$$
\begin{aligned}
&^{2}\mathrm{H} + {}^{1}\mathrm{H} \rightarrow {}^{3}\mathrm{He} + \gamma,\\
&^{3}\mathrm{He} + \mathrm{n} \rightarrow {}^{4}\mathrm{He} + \gamma \text{ or } {}^{3}\mathrm{He} + {}^{2}\mathrm{H} \rightarrow {}^{4}\mathrm{He} + {}^{1}\mathrm{H},\\
&^{2}\mathrm{H} + \mathrm{n} \rightarrow {}^{3}\mathrm{H} + \gamma,\\
&^{3}\mathrm{H} + {}^{1}\mathrm{H} \rightarrow {}^{4}\mathrm{He} + \gamma \text{ or } {}^{3}\mathrm{H} + {}^{2}\mathrm{H} \rightarrow {}^{4}\mathrm{He} + \mathrm{n}.
\end{aligned}
$$

Let me explain in words what is going on in these reactions. In the first of the two sets of equations (the first two lines), a proton is captured onto a deuteron, making ^{3}He (an isotope of helium that has two protons and one neutron) which is then converted to ^{4}He, either with a neutron capture or in a reaction in which a deuteron adds its neutron to the ^{3}He and releases its proton. In the second set of reactions, a neutron is captured onto a deuteron to make ^{3}H, a triton (an even heavier isotope of hydrogen than the deuteron, since the triton has a proton and two neutrons), which then gets converted to ^{4}He either by capturing a proton or in a reaction in which a deuteron drops off its proton and liberates its neutron. Note that each of these reactions conserves baryons.

That's pretty much all there is to BBN, except that ^{7}Li (with three protons and four neutrons) can be made in tiny amounts by the two reaction series where e^- is an electron and ν_e is an electron neutrino:

$$^{4}\text{He} + {}^{3}\text{H} \rightarrow {}^{7}\text{Li} + \gamma, \text{ or}$$
$$^{4}\text{He} + {}^{3}\text{He} \rightarrow {}^{7}\text{Be} + \gamma, \text{ and then}$$
$$^{7}\text{Be} + e^{-} \rightarrow {}^{7}\text{Li} + \nu_e,$$

On the last line is the reaction by which ^{7}Be, which is not a stable nucleus (it consists of four protons and three neutrons), ultimately decays to ^{7}Li by capturing an electron, as indicated. Neutrinos are very important to our story, but not in the context of BBN. However, for the moment it's worth noting that our Sun emits 1.8×10^{38} (100 trillion trillion trillion) neutrinos per second, 8.4×10^{28} (10,000 trillion trillion) of which impinge on one side of the Earth, and virtually all of them pass right on through. If you close your fist, you'll be enclosing a volume that contains several hundred neutrinos. Of course, it's not the same several hundred neutrinos for very long; the neutrinos will pass through your hand with virtually no recognition of your presence.

Solar neutrinos were the source of one of the major scientific puzzles of the twentieth century, that is, why was the rate of detection of Solar neutrinos about one-third of that predicted by the Standard Solar Model (see the website of the late John Bahcall http://www.sns.ias.edu/~jnb/ and Bahcall et al. [8], for many discussions of the Solar neutrino problem and the standard model). The solution to this puzzle required the efforts of many physicists for several decades, and uncovered a profound aspect of neutrinos, that is, they can change from one type—called flavor—to another. We will have more to say about neutrinos later. Finally, note that neutrinos and electrons are not baryons.

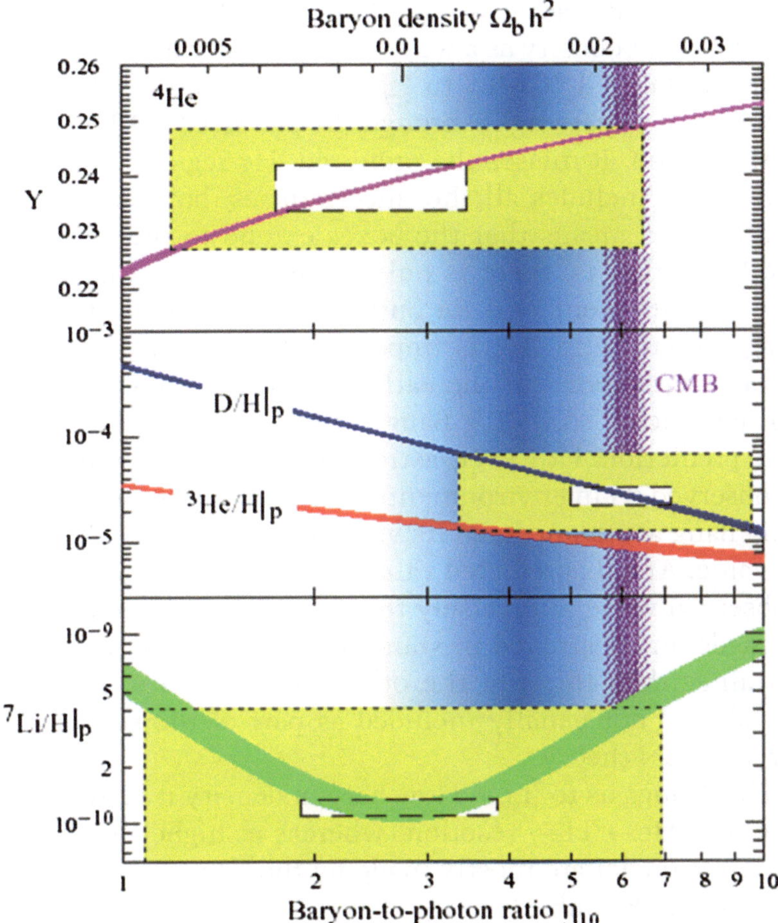

FIGURE 2.5 The abundances of ^{4}He, ^{2}H, ^{3}He, and ^{7}Li as predicted by the standard model of BBN, along with indicated uncertainties in the theoretical predictions. *Boxes* indicate the observed light element abundances (*smaller boxes* 2σ statistical errors, *larger boxes* 2σ statistical and systematic errors added in quadrature). The ^{4}He data are from Fields and Olive [9], the ^{2}H data from Kirkman et al. [10] and Linsky [11], and the ^{7}Li data from Ryan et al. [12] and Pinsonneault et al. [13]. No observations are indicated for ^{3}He, as its primordial value is difficult to obtain. The narrow vertical band along the right side indicates the CMB measured value of the cosmic baryon density. (Reprinted from Boyd [2]. Originally from Fields and Sarkar [14]. Copyright 2004, with permission from Elsevier. Courtesy of Brian Fields)

The abundances of the nuclides made in BBN and the predicted abundances are shown in Figure 2.5, where they are plotted as a function of the "baryon to photon ratio," or baryonic density fraction of the Universe. Prior to WMAP, that ratio was not well known, so it was customary to plot the BBN abundances as a function of that density as a way of determining its value. The vertical region labeled "CMB" in Figure 2.5 gives the WMAP value, which is seen to pass right through the preferred "D/H|p" or ^{2}H to ^{1}H ratio, region. It misses the preferred ^{4}He region, but just overlaps it if one includes all the uncertainties, both statistical and systematic (2σ means that the boxes extend to the 95.4% confidence level, that is, there is only a 4.6% chance that the value of any measurement will lie outside the indicated error boxes). The agreement with ^{4}He is important because nearly all of the neutrons that existed in the early Universe are predicted to end up in ^{4}He nuclei, so its Big Bang value represents the result of a simple prediction. However, astronomers are confident that they have observed ^{2}H in environments that come close to representing the Big Bang abundances, so its value is thought to reflect the Big Bang value. And its predicted value is in excellent agreement with the observed value at precisely the WMAP baryon density. ^{3}He is both made and destroyed in stars, so its BBN value has a greater uncertainty than those of the other BBN nuclides, therefore its BBN value is not usually included as part of the success/failure criteria of BBN theory.

This brings us to ^{7}Li. At low baryon density it is mostly made by the $^{3}\text{H}+{}^{4}\text{He}\rightarrow{}^{7}\text{Li}+\gamma$ reaction, whereas at higher baryonic density mass 7 nuclei are mostly made by the $^{3}\text{He}+{}^{4}\text{He}\rightarrow{}^{7}\text{Be}+\gamma$ reaction, and the ^{7}Be subsequently captures an electron to make ^{7}Li. The mass 7 nuclide production from the former reaction falls off as the baryonic density increases before the latter reaction fully takes over, which is what produces the dip in the ^{7}Li abundance curve. The bottom of the dip is just about what is observed for the Big Bang ^{7}Li abundance. Unfortunately, the CMB value is at a higher baryonic density, and the ^{7}Li abundance at that value is about a factor of 3 above the observed value. The resolution of this discrepancy has been the subject of an enormous amount of research; this is an ongoing research topic for many cosmologists and astronomers.

So how is one to solve the lithium problem? Recent studies [15–17] have looked at possible nuclear reaction solutions, that is, reactions that might contribute to BBN, but which are not well characterized in the BBN computer codes. One particularly interesting aspect of these studies is that, since ^{7}Be is the source of most of the ultimate ^{7}Li abundance during BBN, reactions that might destroy ^{7}Be would mitigate the discrepancy between observation and theory. (Incidentally, there are potentially a lot more reactions that could occur than are indicated above!) All of the above reaction studies identified the $^{7}Be+^{2}H$ reactions as the most promising candidate for a nuclear physics solution, although one of the studies [15] argued that it could not contribute as would be needed in order to reduce the ultimate ^{7}Li abundance by the required factor of 3. The curious feature of this reaction is that it involves a nucleus, ^{7}Be, that has a half-life of 54 days, making it an extremely difficult nucleus on which to study nuclear reactions. However, an experiment has recently been performed with a ^{7}Be beam [18]; the result was that any reactions involving ^{7}Be and ^{2}H could not resolve the lithium problem.

Are there other possibilities that might resolve the problem? One suggestion that has been studied by several authors [19–21] involves the existence of a short lived particle in the early universe that could have become bound to the nuclei that were formed in BBN. The reason that BBN ceases when it does is that the temperature drops to a point at which the nuclides cannot overcome the Coulomb barriers that they must overcome in order to react. (A Coulomb barrier is the electrostatic barrier that exists between two positively charged particles.) However, assuming the short-lived particle was negatively charged, it would reduce the Coulomb barrier as soon as the Universe cooled to the point at which it could be captured into the nuclei that existed at that time, and thus would permit a short resurgence of nucleosynthesis. In so doing, it was found that the ^{7}Li abundance problem could be solved. Of course, if such a particle does exist, it should be produced in the high energy particle accelerators that exist around the world, but has not been seen yet.

To summarize, the current upshot of BBN is that the predicted abundances of ^{2}H and ^{4}He are in good agreement with those observed in stars or other environments that astronomers have identified as representing early Universe abundances. The agreement with ^{7}Li is poor; the predicted value is about a factor of 3 higher than what is observed.

However, we can't make people out of hydrogen and helium, although some politicians are thought to be comprised primarily of gas, but for the rest of us, we need other atoms like carbon and oxygen. These are made in stars, which is the subject of the next chapter.

References

1. M. Hamuy, M.M. Phillips, N.B. Suntzeff, R.N. Schommer, J. Mazo, and R. Aviles, The Morphology of Type IA Supernovae Light Curves, Astron. J. 112, 2398 (1996)
2. R.N. Boyd, *An Introduction to Nuclear Astrophysics*, Univ. Chicago Press, Chicago, 2008
3. S. Perlmutter, G. Aldering, G. Goldhaber, R.A. Knop, P. Nugent, P.G. Castro, S. Deustua, S. Fabbro, A. Goobar, D.E. Groom, I.M. Hook, A.G. Kim, M.Y. Kim, J.C. Lee, N.J. Nunes, R. Pain, C.R. Pennypacker, R. Quimby, C. Lidman, R.S. Ellis, M. Irwin, R.G. McMahon, P. Ruiz-Lapuente, N. Walton, B. Schaefer, B.J. Boyle, A.V. Filippenko, T. Matheson, A.S. Fruchter, N. Panagia, H.J.M. Newberg, W.J. Couch, and The Supernova Cosmology Project, Measurements of the Cosmological Parameters Ω and Λ from 42 High-Redshift Supernovae, Astrophys. J. 517, 565 (1999)
4. G.F. Smoot, C.L. Bennett, A. Kogut, E.L. Wright, J. Aymon, N.W. Boggess, E.S. Cheng, G. de Amici, S. Gulkis, M.G. Hauser, G. Hinshaw, P.D. Jackson, M. Janssen, E. Kaita, T. Kelsall, P. Keegstra, C. Lineweaver, K. Loewenstgein, P. Lubin, J. Mather, S.S. Meyer, S.H. Moseley, T. Murdock, L. Rokke, R.F. Silverberg, L. Tenorio, R. Weiss, and D.T. Wilkinson, Structure in the COBE Differential Microwave Radiometer First-Year Maps, Astrophys. J. 396, L1 (1992)
5. C.L. Bennett, M. Halpern, G. Hinshaw, N. Jarosik, A. Kogut, M. Limon, S.S. Meyer, L. Page, D.N. Spergel, G.S. Tucker, E. Wollack, E.L. Wright, C. Barnes, M.R. Greason, R.S. Hill, E. Komatsu, M.R. Nolta, N. Odegard, H.V. Peiris, L. Verde, and J.L. Weiland, First-Year WILKINSON MICROWAVE ANISOTROPY PROBE (WMAP) Observations: Preliminary Maps and Basic Results, Astrophys. J. Suppl. Series 148, 1 (2003)
6. N. Jarosik, C.L. Bennett, J. Dunkley, B. Gold, M.R. Greason, M. Halpern, R.S. Hill, G. Hinshaw, A. Kogut, E. Komatsu, D. Larson, M. Limon, S.S. Meyer, M.R. Nolta, N. Odegard, L. Page, K.M. Smith, D.N. Spergel, G.S. Tucker, J.L. Weiland, E. Wollack, and E.L. Wright, Seven-Year *WILKINSON MICROWAVE ANISOTROPY PROBE (WMAP)* Observations: Sky Maps, Systematic Errors, and Basic Results, Astrophys. J. Suppl. Series 192, 14 (2011)

7. P.S. Wesson, Olbers's Paradox and the Spectral Intensity of the Extragalactic Background Light, Astrophys. J. 367, 399 (1991)
8. J.N. Bahcall, A.M. Serenelli, and S. Basu, New Solar Opacities, Abundances, Helioseismology, and Neutrino Fluxes. Astrophys. J. 621, L85 (2005)
9. B.D. Fields and K.A. Olive, On the Evolution of Helium in Blue Compact Galaxies, Astrophys. J. 506, 177 (1998)
10. D. Kirkman, D. Tytler, S. Burles, D. Lubin, J.M. O'Meara, QSO 0130-402: A Third QSO Showing a Low Deuterium to Hydrogen Abundance Ratio. Astrophys. J. 529, 655 (2000)
11. J. Linsky, Atomic Deuterium/Hydrogen in the Galaxy. Space Sci. Rev. 106, 49 (2003)
12. S.G. Ryan, T.C. Beers, K.A. Olive, B.D. Fields, and J.E. Norris, Primordial Lithium and Big Bang Nucleosynthesis. Astrophys. J. Lett. 530, 57 (2000)
13. M.H. Pinsonneault, G. Steigman, T.P. Walker, and V.K. Narayanan, Stellar Mixing and the Primordial Lithium Abundance. Astrophys. J. 574, 398 (2002)
14. B.D. Fields and S. Sarker, Big Bang Nucleosynthesis. Phys. Lett. B 592, 1 (2004)
15. R.N. Boyd, C. Brune, G.M. Fuller, and C.J. Smith, New Nuclear Physics for Big Bang Nucleosynthesis, Phys. Rev. D 82, 105005 (2010).
16. N. Chakraborty, B.D. Fields, and K.A. Olive, Resonant Destruction as a Possible Solution to the Cosmological Lithium Problem, arXiv: 1011.0722
17. R.H. Cyburt and M. Pospelov, Resonant Enhancement of Nuclear Reactions as a Possible Solution to the Cosmological Lithium Problem, arXiv:0906.4373 (2009)
18. P.D. O'Malley, D.W. Bardayan, K.Y. Chae, S.H. Ahn, W.A. Peters, M.E. Howard, K.L. Jones, R.L. Kozub, M. Matos, S.T. Pittman, J.A. Cizewski, and M.S. Smith, Phys. Rev. C 84, 042801(R) (2011)
19. C. Bird, K. Koopmans, and M. Pospelov, Primordial Lithium Abundance in Catalyzed Big Bang Nucleosynthesis, Phys. Rev. D 78, 083010 (2008)
20. M. Kusakabe, T. Kajino, R.N. Boyd, T. Yoshida, and G.J. Mathews, Simultaneous Solution to the ^{6}LI and ^{7}Li Big Bang Nucleosynthesis Problems from a Long-Lived Negatively Charged Leptonic Particle, Phys. Rev. D 76, 121301(R) (2007)
21. M. Pospelov and J. Pradler, Big Bang Nucleosynthesis as a Probe of New Physics, Ann. Rev., Nucl. Part. Sci. 60, 539 (2010)

3. What is the Origin of the Rest of the Elements?

Abstract All the nuclei from carbon to lead, and even beyond to thorium and uranium, are produced by nuclear reactions in stars. The reactions that produce each group of nuclei operate under conditions of density and temperature that characterize the phase of stellar evolution in which these reactions occur. Thus helium is made in hydrogen burning, carbon and oxygen are produced in helium burning, neon and more oxygen are made carbon burning, and so forth until the final phase of nucleosynthesis that occurs in massive stars makes iron and nickel. Along the way, during helium burning, heavier nuclei up to bismuth are synthesized. Massive stars ultimately explode as supernovae, which enables another process that produces nuclei as heavy as thorium and uranium. This chapter provides the basic features of all of these processes, as well as some of the basic characteristics of supernovae.

3.1 Introduction to Stellar Nucleosynthesis

Now that we have seen where the hydrogen and most of the helium in the universe came from, we will shift gears and describe how almost all the rest of the nuclides in the "chart of the nuclides" were made. The chart of the nuclides contains all the isotopes of all the elements; it's much more involved than the periodic table, which contains only the elements.

Just to remind you of some basic facts that you learned in your junior high science class: different isotopes of an element have the number of protons that defines that element, but a different number of neutrons. For instance, helium (He) has eight known

R.N. Boyd, *Stardust, Supernovae and the Molecules of Life: Might We All Be Aliens?*, Astronomers' Universe, DOI 10.1007/978-1-4614-1332-5_3,

isotopes, all of which have two protons, although only ^{3}He and ^{4}He are stable, that is, they have not been observed to undergo radioactive decay. Carbon (C) has 15 known isotopes, from ^{8}C to ^{22}C, of which only ^{12}C and ^{13}C are stable. Isotopes that are radioactive are denoted as radioisotopes or radionuclides. As noted in Chap. 1, ^{14}C is a radioactive form of carbon with a half-life of 5,730 years; this is a particularly useful half-life for determining ages of things that have lived at one time, and therefore contain carbon.

Stellar evolution involves a number of phases, each of which involves a different set of nuclear reactions and produces different nuclides. So to understand how all the nuclides are made, we have to follow through all the phases of stellar evolution. And that doesn't produce all the nuclides; two other processes are needed to synthesize most of the nuclides heavier than iron. So here we go.

The cores of stars are so hot and dense that they can "burn" the nuclei they contain through nuclear reactions to synthesize different nuclei; this process is called stellar nucleosynthesis. Note that when I use the word "burn," I am not talking about chemical reactions, these are nuclear reactions. The nuclear reactions generally occur at much higher temperatures, tens or hundreds of millions Kelvin (in this temperature scale, 10 million Kelvin = 18 Million degrees Fahrenheit, and room temperature is about 300 K), and they usually produce a lot more energy than do chemical reactions. They also proceed much less rapidly at typical stellar temperatures than most of chemical reactions we know from our Earthly experiments; that's why stars live for millions or billions of years!

The first stars that formed had to create heavier nuclei, the carbon, nitrogen, and oxygen around which our story revolves, through a myriad of nuclear reactions. What I'll describe below is what goes on in stars that were formed of ingredients that already contained a lot of elements that were heavier than hydrogen and helium. These stars were made of the stuff that had been expelled from earlier generations of stars (several generations of stars are thought to have produced the stuff of which our Sun is made). The first generation burned the primordial stuff, the primordial hydrogen and helium from the Big Bang, but successive generations had some carbon, nitrogen, and oxygen to work with. So on to the phases of stellar evolution.

3.1.1 Hydrogen Burning

In their first stage of stellar evolution, stars burn hydrogen into helium. As noted above, most of the helium we see today was created in the Big Bang, so the helium abundance has increased only a little as time has gone on due to the helium being produced from hydrogen burning in stars. Massive stars spend about 90% of their lives in this stage of stellar evolution. Stars with masses greater than roughly three times the mass of our Sun will have dominant reactions of hydrogen burning that are different than those that operate in less massive stars.

There are actually two sets of nuclear reactions that describe hydrogen burning in stars: the pp (proton-proton)-chains and the CNO (carbon/nitrogen/oxygen) cycles. In our sun, the pp-chain reactions dominate. Just to give you a flavor of how different these reactions can be, I'll write them down. For lower mass stars like our Sun, the reactions go as:

$$^{1}\mathrm{H} + {}^{1}\mathrm{H} \rightarrow {}^{2}\mathrm{H} + e^{+} + \nu_{e},$$
$$^{2}\mathrm{H} + {}^{1}\mathrm{H} \rightarrow {}^{3}\mathrm{He} + \gamma, \text{ and finally}$$
$$^{3}\mathrm{He} + {}^{3}\mathrm{He} \rightarrow {}^{4}\mathrm{He} + 2\,{}^{1}\mathrm{H}.$$

The nuclei in these reactions were defined in the discussion of Big Bang Nucleosynthesis. The symbols e^{+} and ν_e represent a positron, that is, an antielectron (which has the mass of an electron, but a positive charge), and an electron neutrino, as defined above. Don't worry too much about these terms for the moment; we will talk more about the ones that are important to our story in subsequent chapters, and we will definitely return to the mysteries of neutrinos. They will turn out to be crucial to our story, but not yet, and not at the energies at which most of them are produced in hydrogen burning. Note that in these equations, the number of baryons (that is, the number of protons and neutrons) is the same on the right and left sides of the equations, so the baryon conservation law mentioned in Chap. 2 is satisfied. The charge conservation law is also obeyed; the amount of charge on the left and right sides of the equations is also equal. For example, in the first equation, the two protons on the left hand side make the total charge +2. The deuteron, an isotope of hydrogen, has a charge of +1, so the positron is required to make up the additional charge.

In the first equation above, two protons fuse to form a deuteron, a positron, and an electron neutrino. In the second equation, a proton and a deuteron fuse to produce a ^{3}He nucleus and a gamma ray, which is needed to conserve energy. And so forth. The examples discussed so far have defined all the symbols that we will encounter, so that I'll just write subsequent reactions as equations.

In addition to the reactions indicated above, there are a few more that produce nuclei up to ^{7}Be, but those heavier nuclei tend to get returned to lighter nuclei via subsequent nuclear reactions and decays of the radioactive nuclei produced along the way. I've given you a picture, in Figure 3.1, of the three burning chains that convert hydrogen to helium. The one on the left, the so-called pp-I chain, is the one I described above; the other two are responsible for only a tiny fraction of the Sun's energy output and helium production. It is worth noting that there is a profound difference between the reactions that convert hydrogen into helium in stars and those that operated in BBN. This is a direct result of the neutrons that existed during BBN; neutrons do not exist in stars at a high enough abundance level to produce the abundant nuclei,

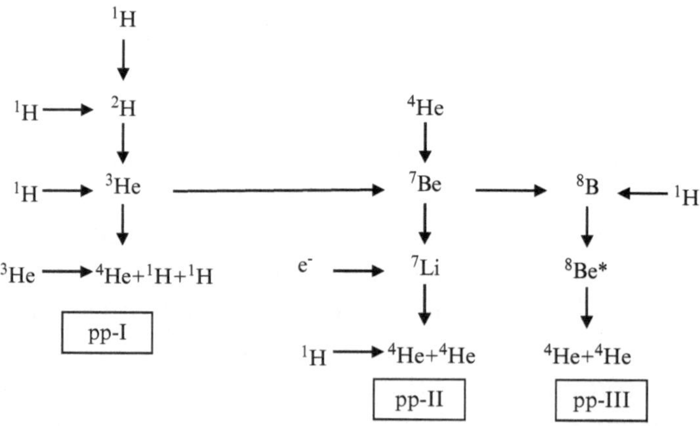

FIGURE 3.1 The three chains that comprise hydrogen burning in the Sun, and in other not-too-massive stars. The pp-I chain produces most of the Sun's energy. The two nuclei involved in each nuclear reaction are indicated by the converging *arrows*, and the resulting nucleus at the point of convergence of the *arrows*. For example, in the chain on the *left*, ^{2}H and ^{1}H combine to form ^{3}He (and emit a gamma ray, not shown in this figure). (From Boyd [1]. Courtesy of University of Chicago Press)

except at the time a massive star explodes and becomes a supernova. So the set of reactions of BBN does not apply to stars.

In each of the three pp-chains shown in Figure 3.1, four protons are fused to produce a ^{4}He nucleus, a particularly tightly bound nucleus (the importance of this will become clearer in a later section of this chapter). If you check the bookkeeping carefully, you'll see that in the pp-I chain, actually six protons go into that chain, but two of them exist in the final reaction along with the ^{4}He nucleus, so the net number of protons used is four. In the pp-II chain, the final equation produces two ^{4}He nuclei, but one of them is an input at the top of that chain, so the net number of ^{4}He nuclei is one. The pp-III chain actually begins with ^{7}Be, which already has one ^{4}He nucleus and three protons included therein, so the one proton that enters that chain produces one net ^{4}He nucleus when the ^{8}Be decays. The asterisk indicates that the ^{8}Be is produced not in its lowest allowed energy state, but the next highest allowed state.

Since we will need to discuss the four identified fundamental interactions that mediate stellar evolution to discuss the CNO cycles, I'll introduce them now. They are, in order of their relative strengths: the strong interaction, the electromagnetic interaction, the weak interaction, and the gravitational interaction. For our present purposes we need only concern ourselves with the first three, and of those, the strong and electromagnetic interactions are much stronger than the weak interaction (it is aptly named).

Getting back to hydrogen burning, more massive stars also convert hydrogen into helium, but by a completely different set of reactions than those of the pp-chains. These reactions, which comprise the CNO cycles, use ^{12}C as a catalyst nucleus. These nuclear reactions look as (and I've also indicated the interaction that mediates each one):

$$^{12}\text{C} + {}^{1}\text{H} \rightarrow {}^{13}\text{N} + \gamma, -\text{(electromagnetic interaction)}$$
$$^{13}\text{N} \rightarrow {}^{13}\text{C} + e^{+} + \nu_{e}, -\text{(weak interaction)}$$
$$^{13}\text{C} + {}^{1}\text{H} \rightarrow {}^{14}\text{N} + \gamma, -\text{(electromagnetic interaction)}$$
$$^{14}\text{N} + {}^{1}\text{H} \rightarrow {}^{15}\text{O} + \gamma, -\text{(electromagnetic interaction)}$$
$$^{15}\text{O} \rightarrow {}^{15}\text{N} + e^{+} + \nu_{e}, -\text{(weak interaction)}$$
$$^{15}\text{N} + {}^{1}\text{H} \rightarrow {}^{12}\text{C} + {}^{4}\text{He}. -\text{(strong interaction)}$$

In these equations, "C" stands for a carbon nucleus, "N" for a nitrogen nucleus, and "O" for an oxygen nucleus. When a gamma ray, γ, is produced, the interaction involved is electromagnetic. When a neutrino, ν, is produced, the interaction is the weak interaction. When there are only nuclei in the equation, the strong interaction mediates the reaction. In this second set of reactions, it is seen that the ^{12}C nucleus (which has six protons and six neutrons) catalyzes the capture of four protons (the ^{1}H), and with two "beta decays," produces a ^{4}He nucleus and returns the original ^{12}C nucleus. Beta decay is the process by which one nucleus decays to another, emitting a positron, the e^{+}, and an electron neutrino, indicated by ν_e. In beta decay, the total number of baryons, protons + neutrons, in the initial nucleus (for example, ^{13}N—seven protons and six neutrons) is the same as that in the final nucleus (in this case, ^{13}C—six protons and seven neutrons). Beta decay proceeds via the "weak interaction," which will turn out to be important for our story, but not just because of hydrogen burning. So it's just another name at this stage. We'll return to our discussion of beta decay in Chap. 7. I've also included a figure to show these reactions pictorially.

As you can see from Figure 3.2, the reactions I gave you above are only one of the CNO cycles, specifically the loop at the left, but the other cycles don't contribute much energy or nucleosynthesis except at the very high temperatures that occur in very massive stars (roughly ten or more solar masses). Like the pp-chains, though, the net result of each of these cycles is the conversion of four protons into a ^{4}He nucleus, although in the CNO cycles this is done by four proton captures and two beta-decays, which differs from the conversion in the pp-chains, in which the beta-decays are rendered unnecessary by the reaction that produces deuterium.

So why does this CNO cycle burning operate in some stars and the pp-chains operate in others? This depends on the temperature of their environment, and that is determined by the mass of the star since more massive stars are hotter. As you recall again from your junior high science courses, charged particles tend to repel each other if they have the same charge, and to attract each other if they have opposite charges. That repulsive force creates the Coulomb barrier, mentioned at the end of Chap. 2. Since all nuclei, from protons to beyond lead, are positively charged, they

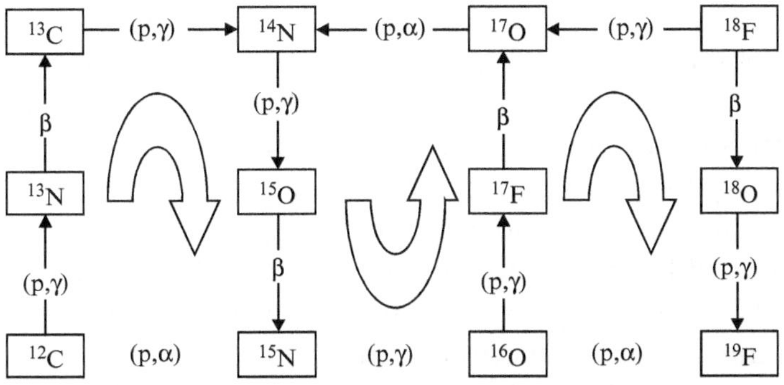

FIGURE 3.2 The CNO cycles of hydrogen burning. In the cycles on the *left* and *right*, the direction of the flow is clockwise, while in the center cycle it is counter clockwise, as indicated. The cycle on the *left*, involving carbon, nitrogen, and oxygen isotopes, is the dominant one, although the cycles in the *middle* and on the *right* can contribute importantly in more massive stars. In these cycles, the beta-decays (indicated by the Greek letter β), generally occur much more rapidly than the nuclear reactions. Nuclear notation is used here: (p,γ) indicates a reaction in which the nucleus, for example, ^{12}C, captures a proton to make a heavier nucleus (^{13}N in this example), emitting a gamma ray, indicated by the Greek letter γ. (Adapted from Boyd [1]. Courtesy of University of Chicago Press)

will tend to repel each other. Nuclei are pretty tiny: around 10^{-14} (one one-hundred-trillionth) meter. In order for them to fuse, for example, for a proton to fuse with a ^{12}C nucleus to make ^{13}N, the proton and ^{12}C nucleus have to get to within that 10^{-14} meter of each other. Although the two positively charged nuclei tend to repel each other, if they have enough energy they will be able to get that close. (For physics aficionados, they can undergo "quantum mechanical tunneling," so they can get close enough to fuse even at relatively low energies.) The temperature of an environment is determined by the motion of the particles in that medium, that is, by their energies, and higher temperature means more energetic particles, so a higher temperature environment will allow its particles to overcome more readily the forces that tend to repel the charged particles. Thus the reactions that operate in stars will proceed more rapidly at higher temperatures.

The ^{12}C nucleus has a lot more positive charge (six units) than a proton (one unit), so the force of repulsion between the proton and ^{12}C will be a lot larger than that between two protons. Thus the first reaction in the pp-I chain, $^1H + ^1H \rightarrow ^2H + e^+ + \nu_e$, will occur at a lower temperature than the first reaction in the main CNO cycle, $p + ^{12}C \rightarrow ^{13}N + \gamma$. And, since less massive stars have lower temperatures than more massive stars, the pp-chain reactions will dominate hydrogen burning in the less massive stars.

But, then, you should ask why isn't pp-chain burning the primary mode of hydrogen burning in all stars? The answer lies with the $^1H + ^1H \rightarrow ^2H + e^+ + \nu_e$ reaction; it is very slow because it is mediated by the weak interaction. Other fusion reactions operate by the electromagnetic or the strong interaction, both of which are much stronger than the weak interaction, so will proceed much faster than the $^1H + ^1H \rightarrow ^2H + e^+ + \nu_e$ reaction, if the temperature is high enough that they can occur. So if a star is sufficiently massive that the temperature becomes high enough for the CNO cycles to operate, they will take over.

One of the nuclei created in hydrogen burning is crucial to our story, and so deserves special attention. This is ^{14}N (7 protons and 7 neutrons). Each of the nuclei in Figure 3.2 is made by one process, for example, ^{13}N is made by proton capture on ^{12}C, and destroyed by another, beta decay to ^{13}C in this example. ^{14}N is the nucleus that is most slowly destroyed in the main hydrogen burning cycle in massive stars, so a lot of it accumulates as the star burns its hydrogen. Therefore, a lot of the ^{14}N is likely to be left over after hydrogen burning ends. Then it will ultimately be expelled into the interstellar medium—the stuff that exists between the stars—following the completion of all the stages of evolution of a massive star, culminated by an explosion in a supernova event. Or, in some stars, the Wolf-Rayet stars mentioned in the introduction, the ^{14}N can be seen in the winds by which the star expels its outer layer or layers. These stars are especially important to our story, as will be explained later on.

The reason that stars can burn nuclei in nuclear reactions is that their temperature and density are very much higher than the densities and temperatures to which we are accustomed. The temperature at the core of our Sun is 15 million Kelvin, or 27 million degrees Fahrenheit. More massive stars burn their hydrogen at temperatures that are 20–40 million Kelvin, or 36 million to 72 million degrees Fahrenheit. The density at the center of the Sun is 150 g/cm^3, about 150 times the density of water, and more massive stars are denser still. So you wouldn't expect these reactions to occur outside of the stars, except in very special situations. And it's not going to be easy to do experiments under conditions that simulate those of the stars!

Of course the hydrogen can't last forever. We believe that our Sun will continue to burn its hydrogen for many more billions of years (it's already been at it for 4.6 billion years). More massive stars, however, consume their nuclear fuel more rapidly; a very massive star may live less than a 100 million years. But it will have several more stages of stellar evolution to complete before it ends its life. And some of those are very important to our story, so let's spend a little time discussing them. To give you a couple of references for our story of stellar evolution, one is the excellent review paper by Woosley et al. [2], the other is a textbook on nuclear astrophysics by Boyd [1].

3.1.2 Helium Burning

When the hydrogen in a star's core has mostly been consumed, the energy that was being produced by the proton induced nuclear reactions ceases. As noted above, the temperature is determined by the motion, or energy, of the particles, and that same energy creates a pressure. Since the energy produced by the nuclear reactions will heat the particles, it will produce the pressure that maintains the size of the region of the star in which the nuclear reactions are occurring for as long as the reactions continue. But when the fuel that burns during that phase is mostly consumed, the reactions cease and the core contracts. As it does so, it gets hotter (if you were wondering, it gets hotter by converting its gravitational potential energy into thermal energy) and denser and, when the temperature becomes high enough, will ignite its next fuel—helium. There will be plenty of helium in the core; it is the "ash" of hydrogen burning. The nuclear reactions that dominate helium burning are pretty simple, and they get us to some very important nuclei, so here they are:

$$^{4}\mathrm{He} + {}^{4}\mathrm{He} \leftrightarrow {}^{8}\mathrm{Be},$$

$$^{8}\mathrm{Be} + {}^{4}\mathrm{He} \rightarrow {}^{12}\mathrm{C} + \gamma,$$

$$^{12}\mathrm{C} + {}^{4}\mathrm{He} \rightarrow {}^{16}\mathrm{O} + \gamma.$$

It took nuclear astrophysicists a long time to figure out the first two reactions, since, as noted above, ^{8}Be only lives for 10^{-16} s. Note that this reaction goes both ways but it will, none the less, build up a tiny abundance of ^{8}Be (some ^{8}Be has to exist for the reaction to go to the left, as shown in the first equation!). The density and temperature are sufficiently high in the helium burning phase that once in a while one of the ^{8}Be nuclei formed for that incredibly short time will capture another ^{4}He nucleus to form ^{12}C. Understanding that reaction even required special ingenuity, as the probability for the ^{12}C to capture the ^{4}He required that a nuclear excited state exist at just the right energy, and with very specific properties, for this reaction to proceed. Protons and neutrons in nuclei can exist in their special quantum mechanical states of existence, just as we found the electrons could in atoms. The ground state is the one with the lowest energy, and thus will be the one in which the nucleus will exist unless a particle, for example, a photon, imparts enough energy to promote it to an excited state. The existence of the crucial state in ^{12}C was guessed by Fred Hoyle (mentioned in Chap. 2) simply because stars do make carbon (and therefore we exist!) long before nuclear physicists were able to perform experiments that showed that such a state did exist. It was ultimately found to be at exactly the energy and with exactly the properties that Hoyle hypothesized.

In any event, this stage of stellar evolution makes two very important nuclei: ^{12}C and ^{16}O, both of which are needed to support our Earthly life. Indeed, both elements may be essential for life in any form to exist. Thus, both will enter our considerations as major players in later chapters.

3.1.3 Subsequent Burning Stages

If a star has sufficient mass, more than eight times the mass of our Sun, it will go through subsequent stages of stellar evolution that next burn the carbon that is made in helium burning; then the neon that is made in carbon burning; then the oxygen that is made in helium burning, carbon burning, and neon burning; and finally the silicon and magnesium that are made in oxygen burning. As the fuel in each stage is consumed, the star contracts and heats up, so that subsequent stages operate in successively hotter and denser environments. In each stage, nuclei are produced that are more tightly configured, that is, their "binding energy" increases,

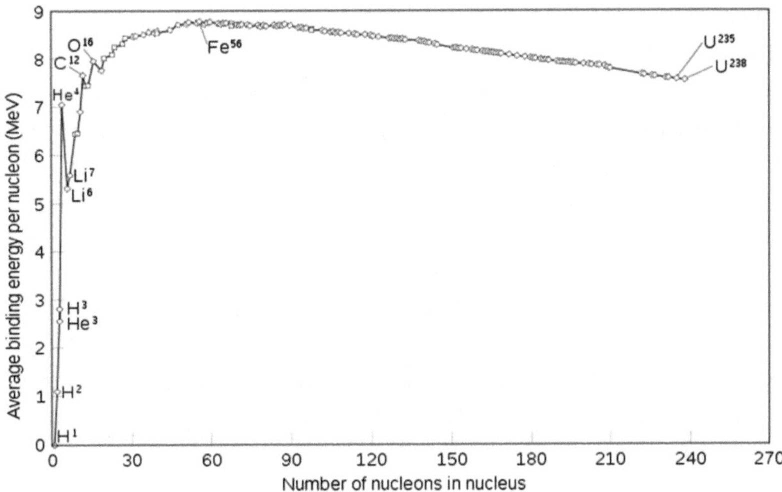

FIGURE 3.3 Binding energy per nucleon of the nuclei in the periodic table. (Courtesy of Wikimedia Commons)

so that net energy is produced by the next set of nuclear reactions. This can be understood qualitatively by a quick inspection of the graph shown in Figure 3.3, which shows the "binding energy per nucleon" for nuclei throughout the periodic table. That is the amount of energy that would be required to completely disassemble a nucleus into its component protons and neutrons divided by the number of protons and neutrons it has. It can be seen that increasing the masses of the nuclei up to iron, Fe, will increase the binding energy per nucleon. Since energy is conserved, when less tightly bound nuclei combine to form a heavier nucleus, net energy will be produced. This is called nuclear fusion.

So let me give you just a few of the details of each of these phases of stellar evolution [1, 2]. Following helium burning the core of a sufficiently massive star will contract and heat up again, and will then undergo carbon burning. The reaction that dominates carbon burning is

$$^{12}\mathrm{C} + ^{12}\mathrm{C} \rightarrow ^{20}\mathrm{Ne} + ^{4}\mathrm{He}.$$

You see this reaction conserves the total number of protons and neutrons, 24, but is not the total fusion of the two ^{12}C nuclei.

That would make ^{24}Mg. But the reaction that makes ^{20}Ne and ^{4}He is just more likely to happen.

It appears that we are just moving up the periodic table in choosing our next fuel, in which case ^{16}O would be the next to burn. But it is a very tightly bound nucleus (see its spike in Figure 3.3), so ^{20}Ne is the next to burn. It just got made; it won't have much time to enjoy its existence! It is destroyed primarily by the reaction:

$$^{20}\mathrm{Ne} + {}^{20}\mathrm{Ne} \rightarrow {}^{16}\mathrm{O} + {}^{24}\mathrm{Mg}.$$

So now, finally, we get to burn the ^{16}O. If it fused, it would make ^{32}S. But once again, that's not the most probable reaction. What actually happens is

$$^{16}\mathrm{O} + {}^{16}\mathrm{O} \rightarrow {}^{28}\mathrm{Si} + {}^{4}\mathrm{He}.$$

What follows oxygen burning is called "silicon burning," or perhaps a more appropriate name would be "silicon melting." Actually both things happen. The temperature of the stellar core has risen to such a high value, several billion Kelvin, by the succession of contractions and heatings that there are many highly energetic photons that can interact with ^{28}Si to produce ^{24}Mg and a ^{4}He nucleus. And more photons to interact with ^{24}Mg to produce ^{20}Ne and a ^{4}He nucleus. And lots of other reactions too. But those light particles, the ^{4}He nuclei, can get captured on other ^{28}Si nuclei to make ^{32}S. And then ^{36}Ar. And so forth. Actually that's a gross oversimplification; the reactions that occur are much more complicated than that, and they involve other light particles, the protons and neutrons. This has to be described by huge computer codes running on huge computers [2]. But the net result of silicon burning is that it destroys some ^{28}Si nuclei and promotes others until they reach the iron-nickel region.

Now look back at Figure 3.3 again. All the reactions we discussed were able to occur because they fused, sort of anyway, two lighter nuclei to make a heavier nucleus, which had a greater binding energy per nucleon, and so the reaction was exothermic—it produced energy. But once you reach the iron-nickel region, making any heavier nuclei by this type of reaction will cost you energy, because those reactions are endothermic, that is, the binding energy per nucleon decreases with increasing mass. So that's about as far as these phases of nucleosynthesis can go.

3.2 After Stellar Burning

So the nuclei around iron and nickel are the primary "ashes" of silicon burning, making that the last stage of burning for a massive star, that is, the iron and nickel cannot be burned into more tightly bound nuclei. So no further fusion process can produce more energy and stabilize the star. This means that the increased energy production that resulted from the contraction following the consumption of each fuel cannot result from burning iron and nickel, and the core of the star collapses, nearly in free fall. The collapse results in very high temperatures; they can increase to several tens of billions Kelvin. At these temperatures the iron and nickel that were produced in the core will be destroyed by the hot bath of photons, the electromagnetic radiation that will accompany that temperature. So all the wonderful nuclei that the star spent its entire life synthesizing, in the core anyway, will be destroyed in a few seconds.

The core of such a star will collapse until it exceeds the density of an atomic nucleus—2×10^{14} grams per cubic centimeter—or a density such that a thimble full of such matter would weigh 200 million tons! The core of the star will become either a neutron star—a star composed primarily of neutrons (with a small fraction of protons, and maybe some even more interesting stuff) or a black hole. Let me put a neutron star into a bit of perspective for you. A neutron star has a mass somewhat greater than that of our Sun, but a radius of about 10 km (6 miles). This is about the size of the beltway around many cities in the United States, and is also the size of the Yamanote Line, the commuter rail system that encircles Tokyo. A black hole, of course, is the ultimate sink. Once you pass the black hole's "event horizon" you'll embark on a very unpleasant journey from which you'll never return. It will be a pretty unpleasant trip, but at least it will be quick. The part of you that extends toward the black hole will feel a much stronger force than the part that is away from it. So, if you go in feet first, your ankles will be stretched more than your neck, so your feet will be pulled off before ... well, you get the picture. However, before the collapse to either the neutron star or the black hole can happen, the star has to cool its core by emitting a lot of energy. It does so within a few seconds by means of the neutrinos mentioned previously, only these are generally at considerably higher energy than those emitted in stellar hydrogen burning. We will return to these neutrinos in a subsequent chapter; they are crucial to our story.

3.2.1 Creating a Core-Collapse Supernova Explosion

When the core of the star collapses, it does so at such speed that it briefly overshoots the density that a neutron star can sustain. This will produce a "bounce," similar to the bounce that a rubber ball exhibits when it hits the floor, only in three dimensions, which will drive an outward going shock wave that will expel the outer regions of the star into the interstellar medium. At least that's how astrophysicists describe the supernova explosion mechanism when they don't have to be too precise. Although the actual explosion mechanism of such stars is not well understood at present, it is clear from astronomical observations that such stars do explode. So even though theoretical astrophysicists haven't yet figured out exactly how that proceeds, nature has solved the problem, at least for many cases.

This explosion will drive the outer layers of the star into the interstellar medium, at least some of the time. The scenario I described above outlined the time evolution of the core of the star, but it also describes the spatial distribution of the different burning stages in a star, which exist in an "onion skin" like structure. The inner parts are hotter and denser than the outer parts, so when the core is undergoing silicon burning, the next shell out is doing oxygen burning, and so forth, so the outer layers are burning helium and finally hydrogen. When the star explodes, expulsion of the outer layers enriches the interstellar medium in the carbon, oxygen, nitrogen, and many other nuclides that are made in the star. This same starstuff will form the constituents of future stars and planets, and of people: we really are composed of stardust!

I've included a picture of a star, in Figure 3.4, at that point in its life just prior to its final stage of core collapse and explosion as a supernova to help you visualize what goes on. The onion skin structure is certainly not something that one should take too seriously; in the olden days when one could only do one

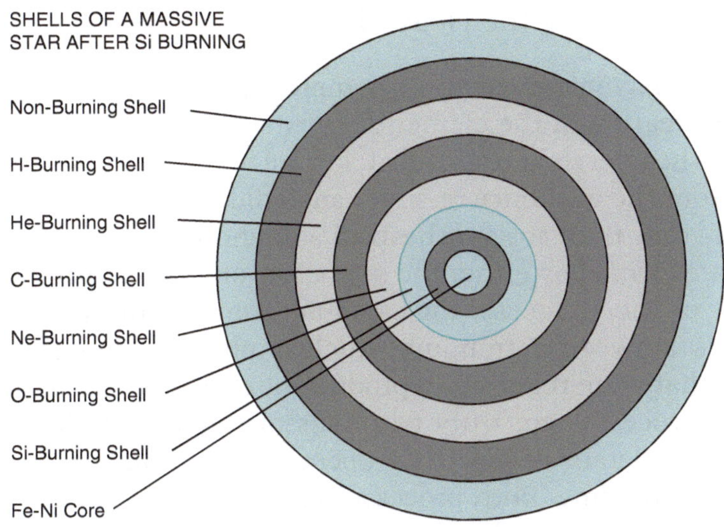

FIGURE 3.4 The burning stages of a well evolved massive star, assumed to be initially at least eight solar masses, showing its burning layers, in the usual "onion skin" form, just before it undergoes its final stage of core collapse. (From Boyd [1]. Courtesy of University of Chicago Press)

dimensional computations one simply had to assume that stars were spherically symmetric, as is indicated in Figure 3.4. However now that computers are large enough to allow calculations that involve more than one dimension, it is clear that turbulence will produce considerable mixing and distortion of the layers [3].

It's worth noting that apparently not all massive stars that end their lives as neutron stars or black holes actually explode in the conventional sense. These "silent supernovae" apparently emit all the neutrinos that their explosive cousins do, as discussed by Fryer [4], but somehow fail to emit very much if any energy in photons. This may be the result of their first forming a neutron star as the collapse process begins, and then, perhaps because they began as very massive stars, collapsing further to a black hole. The neutrinos will get out; they don't interact much with the matter that they pass through, so they are emitted very quickly. But the photons—the particles of electromagnetic energy, scatter around much more before escaping. So there is a good chance they will get swallowed, along with the matter from which they are scattering, by the black hole before they can escape from the star.

3.2.2 Synthesizing the Heavy Elements

Of course I haven't given you a complete picture of nucleosynthesis—the creation of the elements, because I haven't told you how elements heavier than iron and nickel are made, and you know, certainly from the existence of silver and gold jewelry and of nuclear reactors, that they got synthesized somehow. Most of those elements are formed by either the s-process, or slow-neutron-capture process, or the r-process, the rapid-neutron-capture process. The synergies between astronomy, nuclear physics, and theoretical physics that were required to produce our current understandings of those processes are truly triumphs of science, so I'll give you some of the details of how they operate in the context of our current understanding. Both processes are still active areas of scientific study. Figure 3.5 shows a picture of the abundances that are

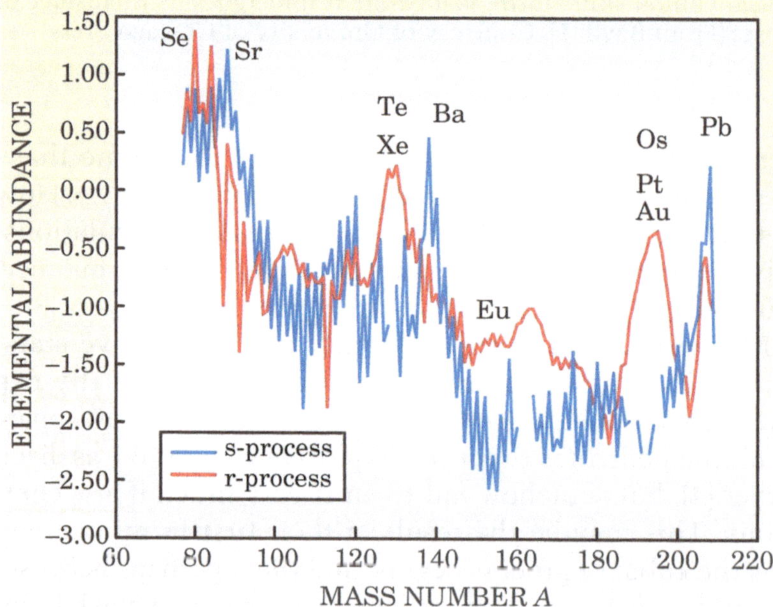

FIGURE 3.5 The abundances (on a logarithmic scale) produced by the s-process and the r-process. The sharp abundance peaks at 88 (Strontium) 138 (Barium) and 208 (Lead) are due to the s-process, and the broader peaks at 130 (Tellurium and Xenon) and 195 (Osmium, Platinum, and Gold) are due to the r-process. (From Cowan and Thielemann [5]. Reprinted with permission of John Cowan and the University of Chicago Press. Copyright 2004, American Institute of Physics)

produced by the two processes, and Figure 3.6 shows the paths of the two processes through part of the chart of the nuclides. Good references for both s- and r-processes are Cowan and Thielemann [5], which focuses on the r-process, but describes a lot about the s-process as well, and of course, Boyd [1].

The s-process occurs mostly during helium burning, and synthesizes about half the nuclei heavier than iron. Its trajectory through the chart of the nuclides passes along the neutron-rich edge of the stable nuclei, as can be seen in Figure 3.6. Because the neutron captures occur slowly, there is usually plenty of time for any resulting unstable nuclei to undergo beta-decay before their next neutron capture, except for a few interesting cases. Equally important is the fact that, because most of the nuclei involved in the s-process are stable or very long-lived, we can study the neutron capture probabilities on them in the laboratory. These probabilities

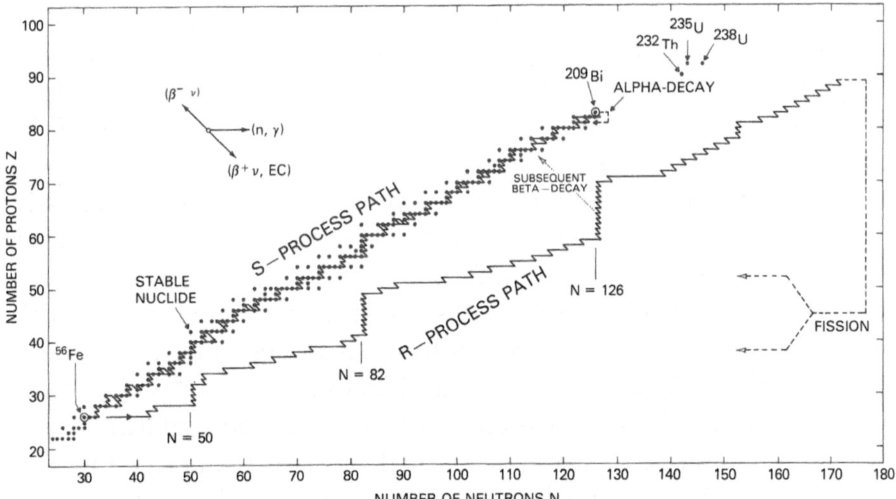

FIGURE 3.6 Neutron capture path for the r-process through a portion of the chart of the nuclides. The nuclides in the band from ^{56}Fe to ^{209}Bi are stable or very long-lived nuclides. Thus the r-process path is seen to proceed 10–30 neutrons beyond stability slowing at the neutron closed shells at 82 and 126 neutrons, until the nuclei become so massive that β-delayed fission and neutron induced fission send the very massive nuclides back to much lighter ones. After the r-process ends, the nuclei that congregated at the N = 82 and 126 neutron closed shells β-decay back to the stable nuclides [6], thus creating the r-process peaks. The r-process path was computed [7] for the conditions $T_9 = 1.0$ and neutron density = 10^{24} neutrons per cubic centimeter. (From Boyd [1], adapted from Rolfs and Rodney [8]. Courtesy of Claus Rolfs, and of the University of Chicago Press)

are the dominant unknowns in the s-process, so that means we can calculate most of the s-process abundances to fairly high accuracy. This will not be the case for the r-process, which we'll discuss shortly.

So, you might ask, how do we know what the abundances are for these two processes? We talked a little bit about astronomy earlier in this Chapter, and we will have more to say about that in Chap. 4. But for the time being, please just accept that astronomers can identify elements that are in the periphery of a star from the light that the star emits, or perhaps the light that is absorbed, in the star's surface. As stars evolve and add their newly synthesized elements to the interstellar medium, the abundances of these elements in the interstellar medium increases. Then as new stars are formed from that material, the abundances of those elements will be larger than they were in the preceding generations of stars. So astronomers can tell how far back in time the stuff that they observe in the periphery of a star was made by seeing how much of key elements, carbon, oxygen, iron, they observe in that star.

So what does that have to do with the abundances of the r-process and s-process, which make the heavier elements? We know that more massive stars go through all their stages of stellar evolution and end their lives more rapidly than less massive stars, so that the products of the nucleosynthesis that are seen in the stars with the lowest abundances of the characteristic elements will have come from the most massive stars. We also know that the most massive stars are responsible for r-process nucleosynthesis. Thus the stars with the lowest abundances are generally thought to have almost totally r-process elements. Indeed, the signatures of the s-process do not seem to appear until later in Galactic time, and this is also consistent with our understanding of these processes.

Ultimately, the stars that are formed out of the stuff of the interstellar medium have both s- and r-process nuclides. Since, as noted above, one can calculate the s-process abundances, one can then subtract those from the total abundances to get the r-process abundances. And, remarkably, these are the abundances that the astronomers have observed in many of the stars with very low abundances of carbon, oxygen, and iron [9, 10].

As we discussed in the chapter on Big Bang Nucleosynthesis, neutrons in isolation are not stable particles, so they have to be captured pretty quickly after they are produced. That will be the case when they are produced in a star; since they have no charge, they don't have to worry about being repelled by the Coulomb barriers of the nuclei into which they would be likely to get captured, so will be captured rather quickly once produced. Therefore, their capture rate is not the rate that makes the s-process the slow-process; it's their rate of production. But that's not all; how the s-process occurs at all is complicated! It has to have stellar regions that are undergoing hydrogen burning in close proximity to regions that are undergoing helium burning in order to produce the neutrons. In the hydrogen burning regions, proton captures produce ^{13}N, which as we discussed above, beta decays to ^{13}C. Now the ^{13}C has to get to a helium burning region so that the reaction:

$$^{13}C + ^{4}He \rightarrow ^{16}O + n$$

can produce the necessary neutrons to run the s-process. The nuclear physics is the easy part of this description; it is the hydrodynamics that allows the hydrogen burning region to produce the ^{13}C, and the helium burning region then to convert that to ^{16}O and a neutron, which makes this tricky. And helium burning doesn't happen the same way during its entire period; one also has to describe the dynamics of that process; its details evolve over the time period during which helium burning operates.

But there's one more bit of chicanery to which Nature has subjected us. In the hydrogen burning zones that produce the ^{13}C, the nuclei therein can't be processed for too long, or they will make ^{14}N. That's bad for the s-process; ^{14}N is a "neutron poison," which will consume neutrons via the reaction $^{14}N + n \rightarrow ^{14}C + p$. So the description of the s-process involves huge computers and huge computer codes. And even then the calculations don't always work out. But apparently Mother Nature is cleverer than we are; she knows how to make it work!

As the s-process moves along its path, it will pause when it gets to the neutron closed shells. Neutrons and protons in nuclei have closed shells just as the electrons do in an atom, (remember back to your junior high science class again) and it is more difficult to capture another neutron on a nucleus for which the neutron shell has just closed. As the s-process progression pauses, abundance will build up at those nuclei; this produces the s-process abundance peaks at Strontium, Barium, and Lead. If the s-process moved more rapidly those peak locations would be shifted. But because the s-process is slow, those peaks occur right at the neutron-closed-shell nuclei.

The heaviest nuclide made by the s-process is ^{209}Bi. What happens if ^{209}Bi captures another neutron is that it beta decays to ^{210}Po, and then expels a ^{4}He nucleus and ends up back at ^{206}Pb. Sometimes you just can't win.

However, we know that heavier nuclides do get made somehow. They are all unstable, but some of them live so long (examples are uranium and thorium) that we know they are synthesized. We also know that whatever process synthesizes them must act quickly; that would circumvent the problem with getting past ^{209}Bi since, if one can capture another neutron on ^{210}Bi before it has time to beta decay, one can get to heavier nuclei. So this gets us to the r-process. There's one other fascinating aspect of the r-process, which is that it is "primary." What that means is that, unlike the s-process, which processes the preexisting "seed" nuclei and just boosts them to somewhat heavier nuclei, the r-process always seems to start with the same seeds, no matter how many heavy nuclei already existed in the star before the r-process began. Indeed, it appears that the r-process destroys all preexisting nuclei and starts from very basic constituents: protons, neutrons, and ^{4}He nuclei, each time it occurs.

There isn't complete agreement as to where the r-process occurs, but the favored site is core-collapse supernovae. At that time there appears to be an intense neutrino wind that blows out from the center of the star, and this produces a high neutron density region. Both seem to be essential for the r-process to work. The maximum temperature that is achieved as the core collapse goes through its successive stages is several tens of billions Kelvin. At that temperature all the preexisting nuclei are destroyed. As the star cools, the seed nuclei for the r-process form, making nuclei up to roughly mass 100. The r-process then converts those seed nuclei to all the nuclides it forms, all the way to uranium and plutonium and even beyond, in seconds, passing through nuclei scientists have never been able to make, even in our most sophisticated

Earthly laboratories. But, once again, Mother Nature doesn't seem to have any problem producing them.

As suggested in Figure 3.6, the r-process moves along a trajectory that is perhaps 20 neutrons to the neutron-rich side of stability. However, it will also pause at the neutron closed shells that occur along that pathway, and some abundance will build up at those points. This produces the r-process abundance peaks at mass 130 and 195. When the r-process has ended, those neutron-rich nuclei will beta-decay back to stability. Thus the stable nuclei at those peaks are very different from the ones that the r-process encountered along its path; they are much less neutron-rich than the ones along the r-process path. However, those peaks are unquestionably the signatures that the r-process has occurred.

You might ask why there isn't a "medium-neutron-capture process," which progressed at a rate somewhere between the s-process and the r-process. Or perhaps your question would be why don't we think such a thing occurs. The s-process abundance peaks show that there is a process that moves slowly through the chart of the nuclides, and the abundances of elements like uranium and thorium show that there is indeed a very rapid process, and the r-process peaks even tell us which are the very neutron-rich nuclei through which the r-process passed. If there were a medium-process, we would see peaks associated with it. There aren't any such peaks. Besides, there is no stage of a star's life that we know of that would produce a medium process. So we have only a slow-process and a rapid-process.

3.2.3 White Dwarfs

I described what happens to massive stars, but not to stars with masses less than eight solar masses. These will not have sufficient mass to complete all their stages of burning, so will not ever become supernovae of the type described above, that is, core collapse supernovae. They may complete some of those stages, and ultimately assume the gentlemanly retirement status of "white dwarfs," stars that are composed either of carbon and oxygen, or of magnesium and neon (depending on how many evolutionary stages they were able to complete), and have masses of not more than around 1.4 times the mass of the Sun. The rest of their mass will be shed in "stellar winds," which will enrich the interstellar medium in the elements that were synthesized in the outer shells of the star. A white dwarf will be the final fate of our Sun. These stellar cinders slowly cool in time, but won't perform any additional nucleosynthesis to contribute to the interstellar medium.

White dwarfs are essentially huge atoms. Although they have masses of about that of our Sun, they are about the size of the Earth. Their pressure is maintained by the electrons they contain—through so called "electron degeneracy pressure"—which is the same pressure that maintains the sizes of atoms. This is a result of a principle of "quantum mechanics," called the "Pauli Principle," which only allows one electron to occupy each quantum state. The state of existence of every electron in an atom, or in the white dwarf, is specified, and no other electron will have the same specifications in either object. So when you fill up all the electron quantum states of something that is the mass of the Sun, you get something that is about the size of the Earth. However, there is a maximum mass for a star that can be supported by electron degeneracy pressure; that was shown many years ago by Chandrasekhar. But this is not essential to our story, so I won't tell you any more of the details. But look in Boyd [1] if you want to know more, and are not intimidated by quantum mechanics.

3.2.4 Type Ia Supernovae

However, if the white dwarf happens to be a member of a pair of stars, it may not just sit there for eternity. In that case, the white dwarf may attract matter from its companion, and that will permit it to exceed the maximum mass that a white dwarf can have. In that case, the electron degeneracy pressure will be overcome, the star will undergo thermonuclear runaway, and the entire star will explode into the interstellar medium. There will be no remnant, no neutron star or black hole. This explosion is also a supernova, but is called a Type Ia supernova; these were used to measure distance scales in the cosmology observations discussed in Chap. 2. Since many white dwarfs are composed of carbon and oxygen, their explosion will also expel those elements on which we depend for life into the interstellar medium. But they won't produce nitrogen; that's restricted to hydrogen burning, and these stars have very little hydrogen. They also produce very few neutrinos; those come only from the core-collapse supernovae. And the model that is the primary one discussed in this book for production of amino acids needs neutrinos to succeed. All the models of amino acid production need nitrogen since every amino acid contains nitrogen. So Type Ia supernovae just won't provide the ingredients that we require.

Before leaving the subject of supernovae, we should just note that a core-collapse supernova emits of order 10^{53} ergs of neutrinos, or about 10^{57} neutrinos, in about 10 s (an erg is a tiny unit of energy by human standards; a paper airplane in flight would have about 10^5, or 100,000, ergs of energy). Although an erg is a small unit of energy, 10^{53} of them is more energy than our Sun will emit in its entire lifetime. That's all but a small fraction of the energy stored in gravitational energy in the core of the supernova at its formation; the rest, of order 1%, goes into the light emitted and the shock wave that explodes the star. When a supernova occurs it can outshine all the other stars in its galaxy for several months from its photons, and that's a tiny fraction of its total energy output! We'll come back to the extraordinary properties of supernovae in a subsequent chapter.

References

1. R.N. Boyd, *An Introduction to Nuclear Astrophysics*, Univ. Chicago Press, Chicago, 2008
2. S.E. Woosley, A. Heger, and T.A. Weaver, Evolution and Explosion of Massive Stars, Rev. Mod. Phys. 74, 1015 (2002)
3. W.D. Arnett and C. Meakin, Toward Realistic Progenitors of Core-Collapse Supernovae, Astrophys. J. 733, 78 (2011)
4. C. Fryer, Neutrinos from Fallback onto Newly Formed Neutron Stars, Astrophys. J. 699, 409 (2009)
5. J.J. Cowan and F.-K. Thielemann, r-Process Nucleosynthesis in Supernovae, Phys. Today, 57, 47 (2004)
6. F.-K. Thielemann, J. Metzinger, and V. Klapdor, Beta Delayed Fission and Neutron Emission: Consequences for the Astrophysical r-Process and the Age of the Galaxy, Z. Phys. A 309, 301 (1983)
7. P.A. Seeger, W.A. Fowler, and D.D. Clayton, Nucleosynthesis of Heavy Elements by Neutron Capture, Astrophys. J. Suppl. Ser. 11, 121 (1965)

8. C. Rolfs and W.S. Rodney, *Cauldrons in the Cosmos*, University of Chicago, Chicago (1988)
9. J.J. Cowan, A. McWilliam, C. Sneden, and D.L. Burris, The Thorium Chronometer in CS 22892–052: Estimates of the Age of the Galaxy, Astrophys. J. 480, 246 (1997)
10. I.U. Roederer, J.J. Cowan, A.I. Karakas, K.-L. Kratz, M. Lugaro, J. Simerer, K. Farouqi, and C. Sneden, The Ubiquity of the r-Process, Astrophys. J. 724, 975 (2010)

4. Creation of Molecules in the Interstellar Medium

Abstract Most of the information we have about the cosmos has come to us in the form of electromagnetic radiation—photons—ranging in energy from the very energetic gamma rays through visible light and extending to radio waves. Only a small fraction of the photons in the electromagnetic spectrum can actually penetrate the Earth's atmosphere, and this dictates the basic features of the "telescopes" used to detect them. Thus, the telescopes used to detect the photons over much of the electromagnetic spectrum must be put into space. This chapter describes the basics of the electromagnetic spectrum as well as the devices that are used to detect the photons of all energies. It also describes the critically important biomolecular information that has been obtained from meteorites that have come to Earth, as well as the anticipated further information at it is hoped will be obtained from future from space missions to comets. Finally, the Drake equation, which characterizes the possibility that we will receive signals from another civilization, is discussed.

4.1 The Electromagnetic Spectrum

Now that we have seen how the elements are synthesized in stars we need to figure out how molecules are formed in the interstellar medium. There is no doubt that molecules are formed; astronomers have been observing them for many years in the "molecular clouds" in the interstellar medium via the spectra emitted from the molecules in the gas phase, or from interstellar dust grains. Astronomers can analyze the light coming from stars with a "spectrograph," an instrument that allows measurement of the intensity of light at each wavelength. Red light has a larger

R.N. Boyd, *Stardust, Supernovae and the Molecules of Life: Might We All Be Aliens?*, Astronomers' Universe, DOI 10.1007/978-1-4614-1332-5_4,

wavelength than yellow light, which has a larger wavelength than blue light, etc. But the colors that we know represent only a tiny component of the electromagnetic spectrum, which ranges from the (very high energy) gamma rays, through X-rays, then to ultraviolet light, and on to visible light, then to the infrared, then to radio waves; these are all electromagnetic radiation, and the "particles of electromagnetic radiation" are photons.

Very high energy gamma rays are produced by high energy processes in the cosmos, or in the Earth's atmosphere by some high energy nuclear reactions. Somewhat lower energy gamma rays can be produced by some of the same processes as those that produce the highest energy gamma rays. But they can also be produced by transitions between allowed energy levels in nuclei that have gained energy from a previous interaction to be in an "excited nuclear state," either in the cosmos or in the Earth's atmosphere. X-rays can be produced by interactions between particles in the Earth's atmosphere (the x-ray machines used by your doctor or dentist utilize these same processes), or by transitions between allowed atomic states in heavier nuclei that have been promoted to an excited state by a previous interaction.

The protons and neutrons in an atomic nucleus can be arranged in many different ways, and each will have a total energy associated with it that is produced by the forces (strong, weak, and electromagnetic) between the particles. However, not all configurations that one might conjure up are allowed; nuclei are microscopic systems, and as such are governed by the laws of quantum mechanics (on which I won't dwell, probably to your everlasting gratitude). As noted earlier, the lowest allowed total energy configuration is called the "ground state," while other allowed configurations are excited states. Most nuclei have many excited states, which can decay to an energetically lower lying state, usually with the emission of a gamma ray to conserve energy.

The same general quantum mechanical principles apply to the electrons in an atom, although the configurations of the atomic electrons are governed primarily by the attractive Coulomb force between the (negatively charged) electrons and the (positively charged) nucleus, and by the repulsive Coulomb interaction between the electrons themselves. Atoms will also have a lowest energy state, a ground state, and other allowed electronic configurations that will produce a variety of excited states. Decays of the atomic excited states to lower energy states will produce x-rays, ultraviolet light, or even visible photons, unlike the decays of nuclear excited states, which produce the much more energetic gamma rays.

As can be seen in Figure 4.1, all of the gamma rays, X-rays, and ultraviolet photons are sufficiently energetic that they will be absorbed as they pass through the Earth's atmosphere. This means that if they are to be detected directly they must be observed in space borne detectors. Several such detectors have provided a wealth of data on astrophysical objects. The highest energy gamma rays, in interacting with the Earth's atmosphere as they approach Earth's surface, do produce observable signals from those interactions. Several ground-based observatories have been built to observe gamma rays by detecting the secondary particles they produce when they interact with the atoms in the Earth's upper atmosphere. Suffice it to say that these various observatories have allowed astronomers to study some of the most energetic processes that occur in our Universe.

As can also be seen in Figure 4.1, the Earth's atmosphere becomes transparent to photons in the very narrow window of

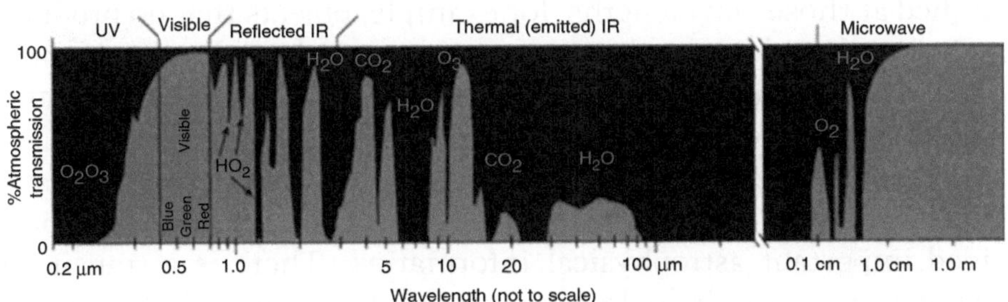

FIGURE 4.1 Transmission (*gray*)/absorption (*black*) of electromagnetic radiation by the Earth's atmosphere as a function of the wavelength of the radiation. Note the rather limited region that allows visible light to be transmitted to the Earth's surface. The other region in which transmission occurs is that of radio waves. The molecules responsible for absorption at various wavelengths are as indicated. Various NASA space observatories, including (from the most energetic to the least) Fermi (high energy gamma rays), Chandra (X-rays), Hubble (visible), Spitzer (infrared), and the National Radio Astronomy Observatories, span the electromagnetic spectrum. The gamma-ray part of the spectrum extends well to the *left* of the graph. (Courtesy of NASA and NASA's Earth Observatory)

visible wavelengths. Surely evolution guaranteed the coincidence between this opacity window and the wavelengths that constitute "visible" light! This is the reason that the optical telescopes that began with Galileo, and have now evolved to levels of sophistication far beyond even his imagination, can provide useful data. However, the Earth's atmosphere, while not absorbing those photons, does still have an effect on them. Fluctuations in the atmosphere make the spatial resolution that one can achieve of much less quality than could be attained without the atmosphere. This is the reason that optical telescopes, the Hubble Space Telescope, and others, were put in space. However, "adaptive optics," a technological development that allows a tuning of the shape of the telescope mirror to refine its focus on a short timescale, together with the use of either real or artificial guide stars, have pushed the capabilities of ground based telescopes to nearly the same level of resolution as the space borne telescopes. These many developments have allowed astronomers to study the evolution of stars, as well as the composition of many objects in the Universe.

Between the optical wavelength region and the radio wave region, the Earth's atmosphere again becomes completely opaque; this is the infrared, or "thermal" part of the spectrum. However, there are a large number of astrophysical objects that need to be studied at those wavelengths, for example, objects that do produce radiation, but do so at lower temperatures than those that produce optical radiation, so again observatories have been built and launched into space.

Finally, the emissions from space that are in radio wave frequencies, indicated as "microwave" in Figure 4.1, have also provided important astrophysical information. There is a transition between two states in hydrogen atoms that produce a photon with a wavelength of 21 cm; this falls in the radio wave window. Since most of the baryonic matter in the Universe is actually hydrogen, this transition, along with radio wave telescopes, have allowed a mapping of most of the matter in the Galaxy, and even well beyond.

However, spectrographs provide much higher detail than just basic colors. Each element has a distinctive set of wavelengths that characterize the light it emits or absorbs. Thus such spectral analysis allows for measurement of the elemental abundances in

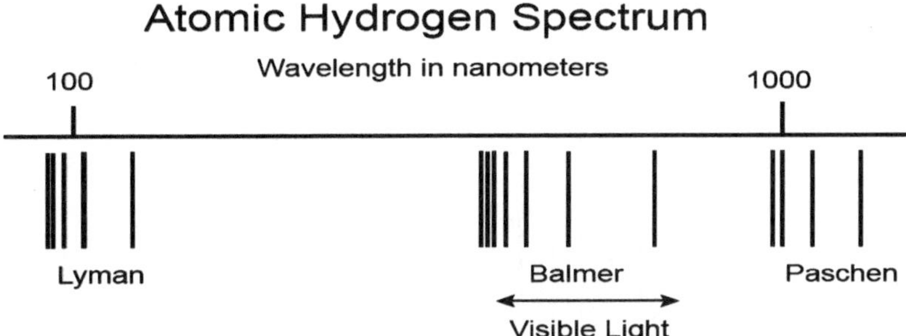

FIGURE 4.2 Emission spectrum for atomic hydrogen. No other atom would exhibit this same set of emission lines.

the periphery of a star. A spectrum for atomic hydrogen (it usually appears in molecular form, that is, in a molecule in which there are two hydrogen atoms) was shown in Chap. 2 to discuss the effects of redshift, but is shown again in Figure 4.2.

Telescopes detect photons, sometimes from emission from atoms or partially ionized atoms (atoms from which one or more of their electrons have been removed), and sometimes from "absorption spectra" from the same entities. In particular, the surfaces of stars are "backlit" by the radiation coming from within the stars; the spectra from that radiation is white, that is, fairly uniformly distributed in wavelength, and exhibiting no characteristic atomic or ionic emission features. The photons from the white spectrum can excite the atoms or ions in the stellar surface, thereby removing those photons from the light emitted from that star at the specific wavelength that characterizes the transition.

Molecules also can be identified by their spectra which, although much more complex than elemental spectra, still provide characteristic signatures. Molecules that have been observed include molecular hydrogen (H_2), water (H_2O), methane (CH_4), carbon monoxide (CO), carbon dioxide (CO_2), ammonia (NH_3), hydrogen cyanide (HCN), formaldehyde (H_2CO), methylacetylene (CH_3CCH), and many others; the body of observations spanning many decades is discussed in the review article by Ehrenfreund et al. [1]. Just so you can see how different molecular spectra can look from atomic spectra, I've included the emission spectrum for carbon monoxide in Figure 4.3.

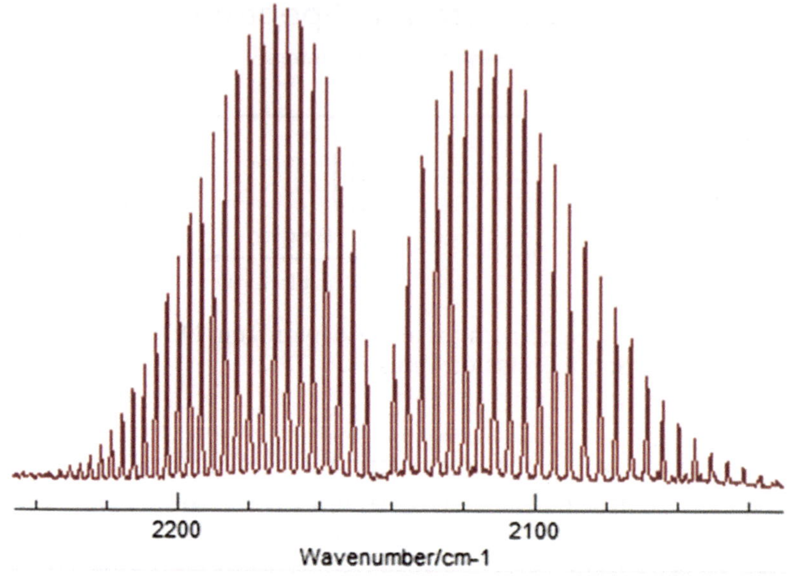

FIGURE 4.3 A section of the (infrared) emission spectrum for carbon monoxide. The x-axis is the "wavenumber," which is 2π divided by the wavelength. As can be seen, the wavelengths associated with molecular spectra are much larger, by roughly a factor of 1,000 (and the energies much smaller), than those that characterize the atomic transitions. The different strengths of the transition lines in this "rotational spectrum" (the different states in which the molecule can exist are the result of its rotation) are determined by the temperature and the intrinsic transition strength. This is considered to be a "simple" spectrum. When there are many different molecules in an emission region, and their spectra are overlapping, it can be extremely difficult to identify exactly which ones are present. Such spectra are referred to pejoratively as "rich". (From http://en.wikipedia.org/wiki/File:Carbon_monoxide_rotational-vibrational_spectrum.png)

4.2 Detecting Photons of Different Energies

As you might expect, rather different "telescopes" are required to view photons over the large spectrum shown (and not shown, for gamma rays) in Figure 4.1; the wavelengths, and hence energies, span many orders of magnitude. The different telescopes need to take into account the ways in which the photons of that energy interact with matter. The highest energy photons—the gamma rays—interact directly with the material they strike to produce strong signals in the detectors. However, they cannot be directed in the same way that visible light can, since they simply penetrate

FIGURE 4.4 "Keck II Laser under moonlight." This shows the twin Keck telescopes, with the laser beam seen that produces the guide star on which the telescope segments are focused. (From Boyd [2]. Courtesy of University of Chicago Press, and of W.M. Keck Observatory)

anything one might try to use to reflect them. X-rays are difficult to redirect, but they can be reflected through small angles. The word "telescope" usually engenders an image of a device with a large dish, now as large as 10 meters, that serves to reflect and focus visible light onto a solid state device that converts the photons into electrical signals. In times past, the photons were recorded on photographic plates, but these could not possibly cope with the rates at which data are acquired with modern telescopes. One telescope, the Keck telescope (see Figure 4.4), utilizes both adaptive optics, which refocuses the separate mirrors of the telescope on time intervals of a fraction of a second, and an artificial guide star in the form of a laser beam that produces an artificial "star" from the light emitted by atoms in the Earth's upper atmosphere that were excited by the laser. The Keck telescope is actually two 10 m diameter telescopes located at an altitude of 13,796 feet (to escape as much of the Earth's atmosphere as possible) on Mauna Kea in Hawaii. Each of the 10 meter primary mirrors is comprised of 36 segments; it is these segments that get refocused on short time scales to optimize the image of the guide star.

Infrared photons require more specialized detectors, but certainly require space-based detection systems because of the total opaqueness of the Earth's atmosphere at those wavelengths. Finally, radio waves can again be detected using Earth based telescopes.

However, other factors need to be considered as well. In order for the telescope to produce a good focus of the photons it receives it must have taken careful account of the precision of its reflective surfaces. For gamma rays this is not a consideration, since the gamma ray will deliver all of its energy to the detector. However, for optical telescopes, the surfaces must be machined to a small fraction of the wavelengths to be detected. Since for optical telescopes the wavelengths are several hundred nanometers, the machining must be good to a small fraction of that: better than 100 nanometer, or one ten thousandth of a millimeter. That has to extend over the ~10 meters of the mirror's surface. Finally, the ability of a telescope to localize the object that it is viewing is determined by the wavelength divided by the diameter of the telescope, so bigger is definitely better.

For radio telescopes the wavelengths are of the order of centimeters, so the required precision is much less taxing. However, radio telescopes need to be huge, hundreds of meters, or even the diameter of the Earth (if one combines the signals from many radio telescopes) in order to localize the object that is being viewed. Radio telescopes are very important to our considerations in this book, so I want to give you a bit more information on them. The radio telescope at the Green Bank Observatory is the world's largest fully steerable radio telescope, and has a primary mirror that is 100 meters in size. A picture of that telescope is shown in Figure 4.5. The one at Arecibo, with a diameter of over 300 meters, is the largest single radio telescope in the world. The Very Long Baseline Array consists of ten radio telescopes, plus sometimes an additional four, ranging in location from Hawaii to Germany. The signals from these telescopes can be combined to produce a telescope with an effective diameter essentially the size of the Earth. We will return to the possible uses and capabilities of radio telescopes in later chapters.

One especially interesting molecule that has been observed on interstellar grains is methanol (CH_3OH). Methanol and ammonia are generally considered to be the keys to chemical reaction

FIGURE 4.5 The Robert C. Byrd Green Bank Telescope, located at the National Radio Astronomy Observatory's site in Green Bank, Pocahontas County, West Virginia. The telescope mirror is 100 meters by 110 meters. This is the world's largest fully steerable radio telescope. (Photo courtesy of the National Radio Astronomy Observatory, Associated Universities, Inc., and the National Science Foundation)

pathways leading to more complex molecules [1], with ammonia [3] being especially critical because of its contains nitrogen. A wide variety of heavier molecules, generally referred to as "polycyclic aromatic hydrocarbons" (PAHs) has also been observed by astronomers. That's a pretty fancy name, but very few of the PAHs have anything to do with our story. Unfortunately, there has not yet been a convincing astronomical observation of any amino acid, which are the molecules in which we have our primary interest, but the limits on their existence are not very stringent; their non-observation may well just be due to sensitivity limits. The molecules that are created tend to freeze out in the icy mantles of the dust grains, so the more volatile species can be frozen into the grains. The temperatures of these grains in the interstellar medium tend to be around 20 Kelvin, 20 degrees above absolute zero (and recall that room temperature is about 300 K on this temperature scale).

4.3 Secrets from Meteorites

While the cosmos is apparently teeming with interesting organic molecules, actually finding the molecules that are relevant to life is a bit trickier. However, one "laboratory" that nature conveniently provides for us, meteorites, makes a completely convincing case that many of the molecules of life are produced in the interstellar medium and that at least some of these can survive the trip to Earth. This last part is not trivial, as meteoroids passing through the Earth's atmosphere (at high speed!) get heated to temperatures that would destroy many molecules.

However, there is one type of meteorite that seems especially able to withstand the high temperatures of Earth entry. These are carbonaceous chondrites, which are chunks of matter containing mostly carbon, and which are capable of maintaining themselves and at least some of the molecules they contain even as they are heated to the high temperatures they achieve upon passing through the Earth's atmosphere. These are thought to be produced in the detritus of the stellar winds of Red Giant stars, among other places; these are stars that have entered their helium burning phase.

One large meteorite fell to Earth on September 28, 1969, near Murchison, Victoria, Australia. The Murchison meteorite fragmented as it entered the Earth's atmosphere, and scattered its pieces over about five square miles. Many of the pieces fragmented again as they hit the ground. However, the pieces that were collected provided us with a stunning potpourri of relevant information. The first study of amino acids from the Murchison meteorite [4] produced the result that the meteorite contained both left- and right-handed amino acids. But the left-handed ones tended to be in greater abundance, although some amino acids were racemic—they had equal numbers of left- and right-handed molecules. A later study [5] concluded that the non-natural (on Earth) amino acids had equal abundances of left- and right-handed components, whereas the naturally occurring ones had been contaminated by Earthly amino acids, so naturally showed a left-handed preference. This result was amplified [6] on the basis that contaminants derived from the ground on which the meteorite had landed had skewed the results. It was noted that another meteorite, the Allende meteorite, had demonstrated amino acid contamination to a significant depth inward from its surface, supporting this contention.

Thus Cronin and Pizzarello [7] did a subsequent study in which they tested the chirality of non-naturally occurring amino acids from the Murchison meteorite. These could not have been contaminated by their Earthly cousins; they don't exist on Earth! What they found was that some of the amino acids had equal amounts of left- and right-handed components, within uncertainties of the measurements, but that others had a definite preference for left-handedness. It might be possible that the high temperatures associated with entry to Earth scrambled the chirality of some of the amino acids, but left more robust ones untouched, or more likely, just not completely scrambled. It would be difficult to argue, however, that the same high temperatures would have produced the chirality observed in some amino acids, given the incredibly tiny energy differences thought to exist between the two chiral states [8–16]. The Cronin-Pizzarello result was confirmed in a subsequent study by Glavin and Dworkin [17]. A segment of the Murchison Meteorite is shown in Figure 4.6.

An additional study by Martins et al. [18] produced a stunning result. It showed that several of the compounds crucial to life other than amino acids also existed in the Murchison meteorite. One of the molecules they detected was uracil, one of the nucleobases (see Chap. 1), a constituent of ribonucleic acid, RNA. To quote their result: "These new results demonstrate that organic compounds [molecules], which are components of the genetic code in modern biochemistry, were already present in the early solar system and may have played a key role in life's origin." Although similar chemicals had been found in the Murchison Meteorite previously, the possibility of contamination was always present, and was difficult to eliminate convincingly. In the study of Martins et al. [18] isotopic ratios of ^{13}C to ^{12}C were measured in the molecules of interest; these ratios were found to be similar to those found in other extraterrestrial samples, but dissimilar to those of terrestrial origin.

Another study was done of meteorites by Callahan et al. [19]; they also found evidence for nucleobases in meteoritic samples. This study was performed on a dozen meteorites from a variety of locations, ranging from Antarctic ice to the Australian location of the Murchison Meteorite. Adenine, one of the nucleobases, was found in eleven of the twelve samples. But more importantly, molecules similar to the nucleobases—so called nucleobase analogs—that are made along with the nucleobases by the chemical processes

FIGURE 4.6 A segment of the Murchison Meteorite, located at the National Museum of Natural History. (From Wikimedia Commons, courtesy of Art Bromage. From http://en.wikipedia.org/wiki/File:Murchison_meteorite.jpg)

thought to occur in the meteorites, were also found. Some of these are found only rarely on Earth, which suggests that both these molecules and the nucleobases found with them are extraterrestrial. Samples of Antarctic ice and of Australian dirt from the location where the Murchison Meteorite hit were also studied. They were found to contain far fewer of the nucleobases and nucleobase analogs than were found in the meteorites.

We will expand on the manner in which the nucleobases form DNA and RNA in Chap. 5, and will comment on the significance of finding the nucleobases in samples from the cosmos in the final chapter.

And there are results from several other meteorites. The Murray meteorite is of the same vintage as the Murchison Meteorite;

it was also found [20] to have amino acids, and that many of them had left-handed chirality. The Allende meteorite [21] fell to earth in 1969 in Mexico, and was also found to contain amino acids. In 2000, a meteorite landed near Tagish Lake in British Columbia, Canada. Analysis of the fragments from this meteorite found a dozen different amino acids. Subsequent analysis, reported by Herd et al. [22], found that different fragments produced different values of chirality, even for the same amino acids. The enantiomeric excess, that is, the percentage by which one chirality exceeded the other, for isovaline was found in one sample to be left handed at a level of a few percent. Perhaps most notable about this meteorite was that some of the amino acid abundances were found to be huge.

Another meteorite that has produced some information about amino acids is the Orgueil meteorite (and it was one of the samples studied by Callahan et al. [19] in their search for nucleobases). Several amino acids were found in that meteorite, and one, isovaline, was found by Glavin and Dworkin [17] to have a significant left-handedness.

4.4 Mining the Comets

Meteorites are certainly an important potential provider of amino acids to Earth, but another potentially important source of extraterrestrial amino acids could be comets. The nuclei of comets have been described as "icy dirtballs" to "dirty iceballs" [1]. They consist of ices, organic compounds, and silicates (inorganic compounds involving silicon). There have been many measurements of cometary constituents, including spectroscopic observations, and satellite fly-throughs of cometary tails.

The molecules observed in comets Hyakutake and Hale-Bopp include much water, carbon monoxide and dioxide, methane, ammonia, and many more complex molecules [1]. A search has been made for glycine, the simplest amino acid, but it has not yet been seen. However, as with the astronomical observations, the upper limit on its existence is not yet very stringent.

The NASA mission Stardust was flown to comet 81P/Wild 2, and at least one amino acid was found in the material which Stardust brought back to Earth. The sample that was returned from this comet was sufficiently large that it was possible to

Figure 4.7 Artist's conception of space mission Hayabusa (literally, peregrine falcon). The large *blue panels* are the solar collectors that provide power to the spacecraft. (From "public domain image." Author: J.R.C. Garry)

measure an abundance for glycine and another amino acid that was determined to be an impurity [23]. The glycine, however, was demonstrated convincingly to be from the comet. Unfortunately, it was not possible to measure any amino acid chirality, as glycine is unique among the amino acids as being non-chiral.

However, we are poised to obtain a huge influx of new information about the constituents of cosmic bodies. One such advance had been hoped from the return of the mission Hayabusa [24–26]. It landed on asteroid Itokawa, and returned to Earth with samples from the asteroid in 2010. It had been hoped that Hayabusa would be able to return with some chunks of the asteroid, but the systems of Hayabusa that were required for this did not work out. However, it may well have dust samples from Itokawa, less material than had been hoped for, but apparently still some asteroidal stuff. This is still being analyzed for a variety of constituents as this book is written, but certainly for any amino acids that are contained in the material that is returned. A simulation of Hayabusa is shown in Figure 4.7.

Additional detailed information should be arriving several years later, if we can be patient (that's not always easy; scientists tend to be pretty impatient!), from the ROSETTA mission ([27]

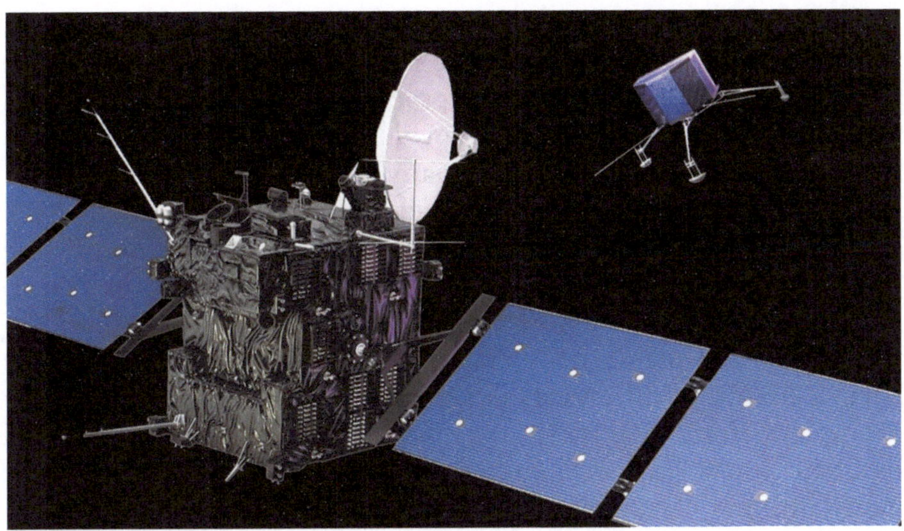

FIGURE 4.8 Artist's conception of the space mission ROSETTA, scheduled to send a lander, Philae lander, shown to the *upper right*, down to comet Churyumov Gerasimenko in 2014. The solar collectors can be seen extending to the *lower left* and *upper right*. (Courtesy of NASA)

and http://www.esa.int/esaMI/Rosetta/SEMYMF374OD_0.html), which is a project of the European Space Agency. ROSETTA is scheduled to land a module on comet Churyumov Gerasimenko in 2014, and remain in orbit around the comet for two years. ROSETTA's lander will have the capability to measure the chirality of the molecules it finds, so should provide yet another critical test of the uniformity of the left-handedness of the amino acids. A simulation of ROSETTA with its lander is shown in Figure 4.8.

Note that virtually every suggested means for creating a dominant chirality in pre-existing amino acids, which are described in detail in Chap. 5, would apply only to the molecules that were very near the surfaces of dust grains or comets. If a processing mechanism such as circularly polarized light, the preferred explanation by many up to now, was found to be the ultimate answer, the chirality of the molecules on the surface of a comet would surely be different from those in the inside (unless they were processed before they became part of the comet, but even that would be a problem if they were initially processed as dust grains of any appreciable size). However, there is another mechanism that can process the molecules, and this would process all of the molecules in any object that existed. This involves the neutrinos, which are emitted in copius amounts from one type of exploding star, so called core-collapse supernovae, and because the neutrinos are so weakly absorbed, they would process the entire grain, or even something as large as a comet or even an unconstrained planet.

It should also be noted that many comets and meteorites may have hit the Earth early in its history, and that each might have carried with it biologically important molecules that could have populated Earth. Indeed, as was discussed in Chap. 1, it is well established that a period of intense meteorite bombardment on the Moon, and presumably also on Earth, occurred early in the Earth's history, and ended about 3.8 billion years ago (see, for example, Strom et al. [28] and references therein). As discussed in Chap. 2, the Universe, and also our Galaxy, were operating for many billions of years before Earth was formed, so there would have been plenty of time for complex molecules to develop and to have the chirality of their amino acids established.

It is interesting to note that primitive life forms are thought to have formed soon (on a cosmic time scale) after the end of the intense meteorite bombardment [29, 30]. However, the first life forms were certainly pretty basic creatures, that is, probably just single celled critters, and even those cells didn't have all the components that our cells have. Development of such sophisticated beasties would require considerably more time. Multicellular organisms came into being about 1.2 billion years ago, and the divergence of plants, animals, and fungi occurred about 960 million years ago. The planet literally exploded with creatures when the Cambrian explosions occurred 540 million years ago. This resulted in the creation of an immense number and variety of living critters, including the many varieties of trilobites that have been found in the fossil record. Some 250 million years ago there was a major setback; a mass extinction occurred at the end of the Permian era. Shortly thereafter, dinosaurs came into being, and ruled the planet until they were wiped out 65 million years ago. Homo erectus didn't appear, though, until about 2 million years ago. We really do represent the youth movement.

Of course, unless the amino acids that were delivered to Earth were all of the same chirality, it would have been difficult for them to evolve into the homochiral environment that currently exists. The comets do have the ingredients to make more amino acids; they have all been found to "contain volatile nitrogen and carbon compounds, in addition to water ice" [1]. And, as noted above, one comet, 81P/Wild 2, was sampled for amino acids by NASA spacecraft Stardust, which returned to Earth in 2006 with at least one amino acid that is believed to have originated with that comet [23]. And we know, as discussed in Sect. 4.3, that at least some meteorites even contain amino acids, a subject to which we will return in the next chapter. So there are plenty of objects in the cosmos that could have brought the molecules of life to Earth, provided they had all been processed to have the same dominant chirality.

The meteorites tell us unequivocally that amino acids, among other biologically interesting molecules, have been created in the interstellar medium, and that they can survive the treacherous journey to Earth. Furthermore, at least some of them have a preferred left-handed chirality. This result is a crucial component to understanding the origin of Earthly amino acids, and to our story.

4.5 The Next Huge Step: Forming Life from the Molecules Delivered to Earth

Obviously we want to take the next step, if we can, from the existence of the primitive molecules of life in outer space, or even as they might be created on Earth, to living beings. This is fraught with uncertainty, since the evolutionary beings that exist now triumphed over their predecessors, thereby eliminating the predecessors. So it is not easy to begin with primitive molecules and Figure out how we got from there to where we are now.

So let's settle for a somewhat simpler procedure for the present. This will not get us to the molecular linkage that we would like, but it is a qualitative time-honored approach to determining the chance that life exists other places in our Galaxy. This goes under the name of the "Drake equation," devised by Frank Drake in 1961. As described by Davies [31], it is less of an equation than "a way to quantify our ignorance." This equation was (as described in Davies book) established to estimate the number of civilizations from which we might detect radio emissions, as that was the focus of searches for extraterrestrial intelligence at that time. The equation gives the number N of possible radio emitting civilizations in our Galaxy as:

$$N = R^{*} F_p N_e F_l F_i F_c L.$$

The different factors are:

R^* is the rate of formation of Sun-like stars in the Galaxy.
F_p is the fraction of those stars that have planets.
N_e is the average number of Earth-like planets in each planetary system.
F_l is the fraction of those planets on which life emerges.
F_i is the fraction of those planets with life that evolves to intelligent life.
F_c is the fraction of those planets on which technological civilization and the ability to communicate emerges.
L is the average lifetime of a communicating civilization.

Some of the estimates of these factors have huge uncertainties, as you will see from our discussion below. In the end we won't really come up with a meaningful estimate for the number of possible radio emitting civilizations in our Galaxy, simply because the uncertainties on some of the factors are so large. The point of discussing the Drake equation, though, is to point out the various factors that are involved in creating a radio emitting civilization.

If life develops on a planet, which is part of a star system, it is assumed that it cannot exist forever, that is, it is born, it lives for a while, then dies. And we will assume that it will not be reborn after it becomes extinct (although if life can be carried to planets from outer space there is no reason why several lineages of life, each perhaps quite different from the previous, could not evolve on a single planet). If our basic assumption about the singularity of life events on each planet is correct, what we need is the stellar formation rate, R^*, not the number of stars in existence at any given time. Furthermore, we need those stars to be somewhat like our Sun; stars that are too small would never ignite to produce the energy to warm their planets, and stars that are too large would incinerate their planets before any interesting life could get started, or the large stars might not live long enough for intelligent life to develop on their planets. Astronomers have given us a good estimate of R^* [31]; about seven Sun-like stars are born per year in our Galaxy.

The known fraction of these stars that have planets, N_p, has grown greatly in recent years due to the planet detecting satellite, the Kepler Space Telescope. Davies gives an estimate of 0.5 for this fraction, but it may now be larger. Even if it were as large as its maximum value of 1.0, though, it would only increase N by a factor of two, and this is an inconsequential uncertainty compared to those associated with other fractions (see below).

The value of N_e, the number of those planets per star that could support life, often referred to as Earth-like planets (although that may be far too restrictive, as is discussed in Chap. 9), is taken by Davies to be two. In our solar system, it is generally felt that Earth and Mars could qualify, although, again as discussed in Chap. 9, the number could be larger when the extreme forms of life are included. In any event, I will assume N_e to be two. Of course, this might require a large "protector planet" to sweep out the space debris that could destroy life with frequent bombardments, a function that Jupiter performs for Earth. So perhaps two is an over estimate when this is taken into account.

Now we get to the really uncertain numbers, starting with F_l, the fraction of Earth-like planets on which life actually evolves. As Davies notes, estimates range from 0.01 to 1.0, a range of a factor of 100! If the molecules of life originate in outer space, and all that is needed is a cosmic stork to deliver them (and given the above discussion about the results of refs. [19] and [20], this may well be the case), then the higher estimate may be relevant. Perhaps it makes sense to assume a (sort of) mean value of 0.1 for F_l. I suspect it is really very close to 1.0.

It is more difficult to estimate F_i, the fraction of planets on which life forms that will also evolve to intelligent life (by Earth-human standards), and those that achieve life that can develop radio communications capability. The values of these quantities are really unknown, and are difficult even to guess with any sort of confidence. This is where it would be valuable to know the intermediate stages of molecular evolution from what is delivered to Earth and what is required for modern human existence. Although we might be able to make some sort of guess as a result of future microbiological experiments, those simply don't exist at present.

The last entity is the length of time that civilizations live. Davies, an admitted optimist, guesses 10,000 years, but others are less optimistic. What terminates a civilization? It could be nuclear war, a natural catastrophe such as a large meteorite encounter, a massive global climate change, or some other form of Armageddon, that interrupts the progress of civilization. Since the proliferation of nuclear weapons seems to pose a huge threat for mankind and some level of global climate change appears to be a reality, 10,000 years seems like a long time to me. I'd opt for 1000 years, but this factor of 10 range (which could be even greater) just shows how difficult it is to make these estimates. A related question is even if a civilization lived for 10,000 years, would it be sending out radio signals for that duration? Communications several centuries down the road may take a very different form from the radio waves that are used at present. These last two entries in the Drake equation make it unrealistic to even try to produce a plausible estimate for the number of possible radio signal emitting civilizations in the Galaxy.

However, a factor that might increase L arises from the possibility that several civilizations might take form, successively, on a single planet. This would have to take into account the time it takes for a civilization to form, and to develop radio communication capability. It has taken our civilization a long time for this to happen, but perhaps we are just slower in our development than most others might be.

Conclusion: even with the large uncertainties on these entities, it seems likely that there are many planets in our Galaxy on which life of some form exists, and even quite a few that could send radio signals. The actual number is obviously difficult to determine, but it seems unlikely, to me at least, that it is zero.

References

1. P. Ehrenfreund, W. Irvine, L. Becker, J. Blank, J.R. Brucato, L. Colangeli, S. Derenne, D. Despois, A. Dutrey, H. Fraaije, A. Lazcano, T. Owen, R. Robert, and an International Space Science Institute ISSI-Team, Astrophysical and Astrochemical Insights into the Origin of Life, Rep. Prog. Phys. 65, 1427 (2002). Quote courtesy of IOP Publishing, Ltd., and of P. Ehrenfreund. DOI: 10.1088/0034-4885/65/10/202
2. R.N. Boyd, *An Introduction to Nuclear Astrophysics*, Univ. Chicago Press, Chicago, 2008
3. Y. Shinnaka, H. Kawakita, H. Kobayashi, E. Jehin, J. Manfroid, D. Hutsemekers, and C. Arpigny, Ortho-to-Para, Abundance, Ratio (OPR) of Ammonia in 15 Comets: OPRs of Ammonia Versus $^{14}N/^{15}N$ Ratios in CN. *Astrophys. J.* 260, 141 (1982)
4. K. Kvenvolden, J. Lawless, K. Pering, E. Peterson, J. Flores, C. Ponnamperuma, I.R. Kaplan, and C. Moore, Evidence for Extraterrestrial Amino-Acids and Hydrocarbons in the Murchison Meteorite, Nature 228, 923 (1970)

5. M.H. Engel and B. Nagy, Distribution and Enantiomeric Composition of Amino Acids in the Murchison Meteorite, Nature 296, 837 (1982)
6. J.L. Bada, J.R. Cronin, M-S. Ho, K.A. Kvenvolden, J.G. Lawless, S.L. Miller, J. Oro, and S. Steinberg, On the Reported Optical Activity of Amino Acids in the Murchison Meteorite, Nature 301, 494 (1983)
7. J.R. Cronin and S. Pizzarello, Enantiomeric Excesses of Meteoritic Amino Acids, Science Magazine 275, 951 (1997)
8. S.F. Mason and G.E. Tranter, The Electroweak Origin of Biomolecular Handedness, Proc. R. Soc. London A397, 45 (1985)
9. S.F. Mason and G.E. Tranter, Energy Inequivalence of Peptide Enantiomers from Parity Non-Conservation, J. Chem. Soc. Chem. Comm. 117 (1983)
10. S.F. Mason and G.E. Tranter, The Parity-Violating Energy Difference Between Enantiomeric Molecules. *Molec. Phys.* 53, 1091 (1984)
11. S.F. Mason, Origins of Biomolecular Handedness, Nature 311, 19 (1984)
12. G.E. Tranter, Parity Violating Energy Differences of Chiral Molecules and the Origin of Biomolecular Chirality, Nature 318, 172 (1985)
13. G.E. Tranter, The Parity Violating Energy Difference Between Enantiomeric Reactions, Chem. Phys. Lett. 115, 286 (1985)
14. G.E. Tranter, The Parity Violating Energy Difference Between the Enantiomers of α-Amino Acids, Chem. Phys. Lett. 120, 93 (1985)
15. G.E. Tranter, Parity Violating Energy Differences and the Origin of Biomolecular Chirality, J. Theor. Biol. 119, 467 (1986)
16. G.E. Tranter, The Enantio-Preferential Stabilization of D-Ribose from Parity Violation, Chem.. Phys. Lett. 135, 279 (1987)
17. D. Glavin and J. Dworkin, Enrichment of the Amino Acid L-isovaline by Aqueous Alternation on CI and CM Meteorite Parent Bodies, Proc. National Acad. Sciences 10.1073/pnas 0811618106 (2009)
18. Z. Martins, O. Botta, M.L. Fogel, M.A. Sephton, D.P. Glavin, J.S. Watson, J.P. Dworkin, A.W. Schwartz, and P. Ehrenfreund, Extraterrestrial Nucleobases in the Murchison Meteorite, Earth and Planetary Science Letters 270, 130 (2008). Use of quote courtesy of Elsevier
19. M.P. Callahan, K.E. Smith, H.J. Cleaves, II, J. Ruzicka, J.C. Stern, D.P. Glavin, C.H. House, and J.P. Dwoprkin, Proc. Nat. Acad. Sci., PNAS Early Edition 13995 (2011)
20. J.G. Lawless, K./a. Kvenvolden, E. Peterson, C. Ponnamperuma, and C. Moore, Amino Acids Indigenous to the Murray Meteorite, Science 173, 626 (1971)
21. http://www.meteoritemarket.com/AMinfo.htm
22. C.D.K. Herd, A. Blinova, D.N. Simkus, Y. Huang, R. Tarozo, C.M. O'D. Alexander, R. Gyngard, L.R. Nittler, G.D. Cody, M.L. Fogel, Y. Kebukawa, A.L.D. Kilcoyne, R.W. Hilts, G.F. Slater, D.P. Glavin,

J.P. Dworkin, M.P. Callahan, J.E. Elsila, B.T. De Gregorio, and R.M. Stroud, Origin and Evolution of Prebiotic Organic Matter As Inferred from the Tagish Lake Meteorite, Science 332, 1304 (2011)

23. J.E. Elsila, D.P. Glavin and J.P. Dworkin, Cometary Glycine Detected in Samples Returned by Stardust, Meteoritics and Planetary Science 44, 1323 (2009)
24. A. Fujiwara, j. Kawaguchi, D.K. Yeomans, M. Abe, T. Mukai, T. Okada, J. Saito, H. Yano, M. Yoshikawa, D.J. Scheeres, O. Barnouin-Jha, A.F. Cheng, H. Demura, R.W. Gaskell, N. Hirata, H. Ikeda, T. Kominato, H. Miyamoto, A.M. Nakamura, R. Nakamura, S. Sasaki, and K. Uesugi, The Rubble-pile Asteroid Itokawa as Observed by Hayabusa, Science 312, 1330 (2006)
25. J. Saito, H. Miyamoto, R. Nakamujra, M. Ishiguro, T. Michikami, A.M. Nakamura, H. Demura, S. Sasaki, N. Hirata, C. Honda, A. Yamamoto, Y. Yokota, T. Fuse, F. Yoshida, D.J. Tholen, R.W. Gaskell, T. Hashimoto, T. Kubota, Y. Higuchi, T. Nakamura, P. Smith, K. Niraoka, T. Honda, S. Kobayashi, M. Furuya, N. Matsumoto, E. Nemoto, A. Yukishita, K. Kitazato, B. Dermawan, A. Sogame, J. Terazono, C. Shinohara, and H. Akiyama, Detailed Images of Asteroid 25143 Itokawa from Hayabusa, Science 312, 1341 (2006)
26. H. Yano, T. Kubota, H. Miyamoto, T. Okada, D. Scheeres, Y. Takagi, K. Yoshida, M. Abe, S. Abe, O. Barnouin-Jha, A. Fujiwara, S. Hasegawa, T. Hashimoto, M. Ishiguro, M. Kato, J. Kawaguchi, T. Mukai, J. Saito, S. Sasaki, and M. Yoshikawa, Touchdown of the Hayabush Spacecraft at the Muses Sea on Itokawa, Science 312, 1350 (2006)
27. W.H.-P. Thiemann and U. Meierhenrich, ESA Mission ROSETTA Will Probe for Chirality of Cometary Amino Acids, Origins of Life and Evolution of the Biosphere 31, 199 (2001)
28. R.G. Strom, R. Malhotra, T. Ito, F. Yoshida, and D.A. Kring, The Origin of Planetary Impactors in the Inner Solar System, Science 309, 1847 (2005)
29. S.A. Wilde, J.W. Valley, W.H. Peck, and C.M. Graham, Evidence from Detrital Zircons for the Existence of Continental Crust and Oceans on Earth 4.4 Gyr Ago, Nature 409, 175 (2001)
30. J.W. Schopf, A.B. Kudryavtsev, D.G. Agresti, T.J. Wdowiak, and A.D. Czaja, Laser-Raman Imagery of Earth's Earliest Fossils, Nature 416, 73 (2002)
31. P. Davies, *The Eerie Silence: Renewing Our Search for Alien Intelligence*, Houghton, Miflin, Harcourt, NY, NY 2010

5. Amino Acids and Chirality

Abstract Amino acids are at the heart of this book, and this chapter discusses them in a general way—for non-experts in organic chemistry. It also discusses linear and circular polarization of light, again for non-experts. This leads to the discussion of the Buckingham effect (Chem Phys Lett 398:1, 2004; Chem Phys 324:111, 2006), by which one can discriminate between left- and right-handed amino acids given the right conditions. A short description of quantum mechanics, again for non-experts, shows how the Buckingham effect would manifest itself in chiral molecules. Then more complex molecules of life—DNA and RNA—are discussed, as well as possible means by which they might be produced once the amino acids have been produced. Finally, evidence is presented which suggests that the usually assumed hostile environment of outer space might provide a nursery for the complex basic molecules.

5.1 A Primer on Amino Acids

So now that we've seen that complex organic molecules are formed in outer space, we need to focus a bit on the main characters in our story, the amino acids. Those amino acids will turn out to have a property, the previously mentioned chirality, that will be the key to sorting through the various scenarios by which the amino acids might have been made, and to understanding where they might have come from. First, a definition: an amino acid is a molecule that has both an amino group, which in chemical symbols is NH_2, that is, one nitrogen atom and two hydrogen atoms, and a carboxyl group, COOH, two oxygen atoms, one carbon atom, and one hydrogen atom. Pictures of two relatively simple amino acids are shown in Figure 5.1—there you can see the two requisite components.

R.N. Boyd, *Stardust, Supernovae and the Molecules of Life: Might We All Be Aliens?*, Astronomers' Universe, DOI 10.1007/978-1-4614-1332-5_5,

The amino acids shown in Figure 5.1 have the carbon atom in the middle attached to both the amino group and the carboxyl group. Such amino acids are called "alpha amino acids," and the central carbon atom is denoted as the α-carbon. These are the amino acids most commonly found in nature. The components of the side chains are what distinguish one amino acid from another. There are other classes of amino acids, but they are not involved so much with the existence of life, so we won't worry about them.

However, this book is mostly about astrophysics, and not organic chemistry. There are other books that focus on the chemistry of the molecules of life, but not on the possible astrophysical aspects. For example, the book by Plaxco and Gross [1] does a superb job of discussing organic chemistry, taking it well beyond the amino acid level to the more complex molecules of life, RNA and DNA. We will discuss the function and structure of those molecules a bit, but leave the gory details to Plaxco and Gross (Plaxco is a biochemist, after all). Thus this book will focus on some aspects of the amino acids that the other books have omitted.

To reiterate a bit, the amino acids are the building blocks of the proteins, which allow our continued existence in a variety of ways. We require 20 amino acids for life. Our bodies can only synthesize about half of them; the others need to be acquired from our food. However, there are many other amino acids—more than one hundred, most of which are not naturally occurring on Earth.

In the famous experiment of the 1950s discussed in Chap. 1, Miller and Urey [2, 3] were able to show that at least some amino acids could be synthesized in a spark discharge in an environment containing water (H_2O), methane (CH_4), ammonia (NH_3), molecular hydrogen (H_2), and very little oxygen. Miller claimed

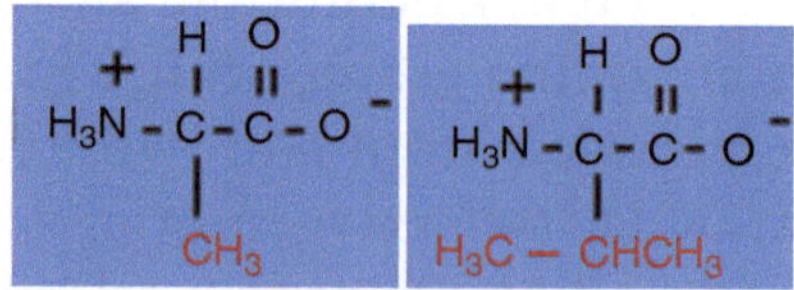

FIGURE 5.1 Structure of alanine (*on the left*) and valine (*on the right*), two amino acids that are essential components of proteins. Note the amino group at the *left*, with an extra hydrogen, and the components of the carboxyl group on the *upper right* on both molecules. (Courtesy of Charles Ophardt, Professor Emeritus, Chemistry, Elmhurst College, Elmhurst, IL)

that five amino acids were produced by this experiment. In 2008, a reanalysis of Miller's archived results [4] showed that 22 amino acids were actually produced in the experiment. This experiment produced the often made claim, attributable many years before to Oparin [5] and Haldane [6], that since the Miller–Urey experiment showed that many of the building blocks of life could grow out of an Earthly lightning storm in some environment that contained the carbon, hydrogen, and nitrogen amino acid constituents, that may be how life, or at least many of the molecules required for life, began. So that's pretty convincing. And this would certainly suggest that we don't need to have anything to do with extraterrestrials, at least as relatives! Our amino acids are home grown, thanks, and if extraterrestrials need amino acids, they can make their own wherever they might be.

5.2 Chirality and Polarization

However, as indicated in Chap. 1, Nature has provided us with an important bit of information that is somewhat difficult to reconcile with the spark discharge creation of the amino acids: that is their chirality. So let's look into that, and try to understand why this amino acid chirality might be difficult to understand in the context of the Miller–Urey creation mechanism.

Chirality refers to the structure of "things"; as noted in Chap. 1, your left and right hands cannot be translated into each other. But when you look at the mirror image of one hand, now they could be translated into each other, at least in principle. So your hands are referred to as having mirror symmetry, since they are mirror images of each other. Similar definitions apply to the structure of molecules, most notably the amino acids. They are also said to have chirality. I've shown, in Figure 5.2, the left-handed and right-handed versions of the amino acid Alanine. There you can see that amino acids and hands enjoy the same sort of chirality.

Chirality can also refer to the direction of rotation of "circularly polarized light" (to be discussed below). The chirality of molecules is closely coupled to the effect the molecules can have on circularly polarized light that shines on the molecules. For instance, when right circularly polarized light shines on a right

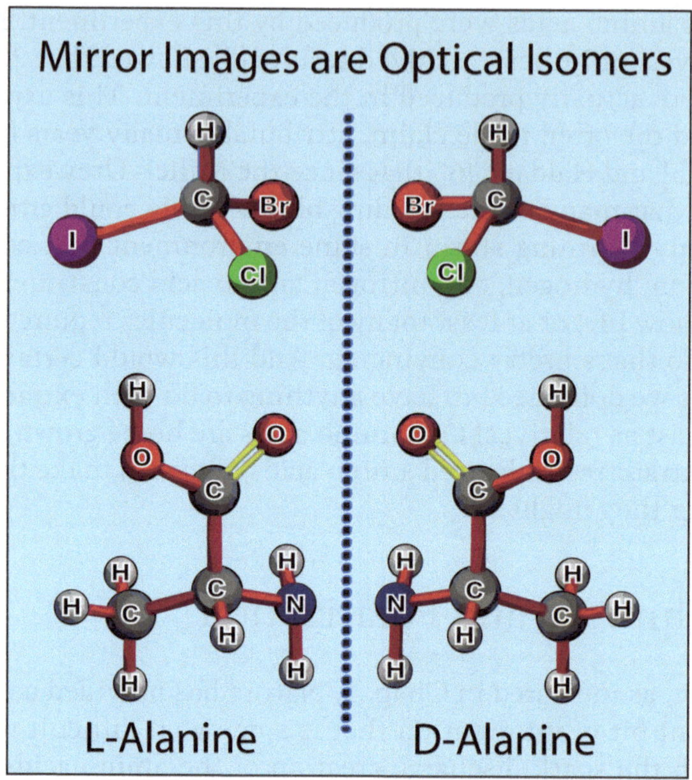

FIGURE 5.2 Mirror images of optical isomers, specifically (*in the lower figure*) L-Alanine and D-Alanine. (Courtesy of Charles Ophardt, Professor Emeritus, Chemistry, Elmhurst College, Elmhurst, IL)

handed amino acid, the light will be affected differently than when right circularly polarized light shines on a left handed amino acid. This effect can take a couple of forms, but one way in which the differences are observed is in the absorption of the light; it will be different for the two polarizations of light as they pass through a chiral medium. This effect has been used to suggest one of the possible ways in which amino acids developed their specific chirality. In Chap. 6 we will talk about some ways to make a collection of racemic molecules (the same number of right and left handed molecules) into an enantiomeric collection (where one chirality is favored); circularly polarized light is one of the ways by which this might happen. In the mean time, to best understand this relationship between circularly polarized light and molecules, we need to talk more about light, some of its specific properties, and how this form of light is affected by these molecules.

Light can be characterized by its "electric field vector." What's an electric field? To give you some feeling for what an electric field is, if there is a charged particle in the region of the electric field, it will produce an acceleration on the particle. What's a vector? Vectors are pictorial representations of quantities that have both direction and amplitude, so we can represent them by arrows of an appropriate length; longer arrows represent stronger electric fields. (Vectors can be used to describe a variety of things, for example, the course of an airplane in flight that is subject to winds.) Incidentally, light also has an associated magnetic field vector, but its direction and amplitude are fixed once the electric field vector and the direction of propagation of the light are fixed. So we will just describe light in terms of the electric field vector.

So let's think of the electric field vector associated with light as an arrow that points in the direction of the electric field while the light particle, the photon, is moving through space. The field will oscillate, in the positive and negative directions, as indicated in the picture in Figure 5.3. That figure assumes that we are looking at the light as if it were frozen in time, that is, as if it were fixed in space as in a snapshot. Thus we see its oscillations as it propagates through space. However, it also oscillates in time; the figure could also represent what we would see if you sat at a fixed location and just watched how the electric field oscillated as the light passed by you. Thus the horizontal axis is labeled as either

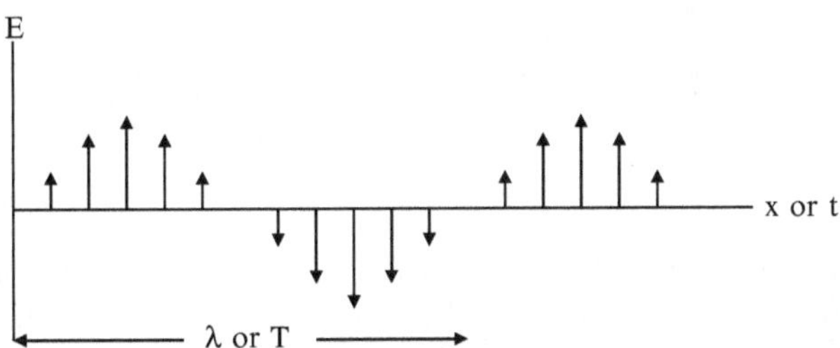

FIGURE 5.3 Electric field vector of a light either as a function of space or of time. If the coordinate on the horizontal axis is space, or x, the length for one complete cycle is called the wavelength, λ. If the coordinate on the horizontal axis is time, the time for one complete cycle is called the period, T.

space or time. The arrows could be oriented so that it points up and down as it oscillates, or it could also be oriented to oscillate left and right, or its oscillations could occur in any other plane. In any of these cases, the light is said to be "linearly polarized".

Now suppose that we added up–down and left–right electric field vectors of the same strength (or length) so that the one going up and down was at its maximum up direction exactly when the one going left and right was maximum to the right. The up–down and left–right electric field components are said to be "in phase" in this configuration; they both achieve their maxima and minima at the same time. Vector addition is pretty basic stuff; one has to maintain both the magnitude of the vector, represented by its length, and its direction (and this requires a little trigonometry). If we do that for this case, then we would end up with an electric field direction that was at 45° to both up and down, and to left and right, as shown in Figure 5.4. Think of someone pushing a box along the floor where they are pushing it to the south, and someone else is pushing just as hard to the east. The motion of the box would then be toward the southeast, it would move a little faster than it would have with either push by itself (but less than twice as fast), and you just performed vector addition. Now assume we are looking in the direction of propagation of the light, that is, we are riding along (if this were possible!) at the speed at which the light is moving—the speed of light. The total electric field resulting from the addition of the two vectors is in the diagonal direction, as shown. This light is also linearly polarized in the up-right and down-left direction.

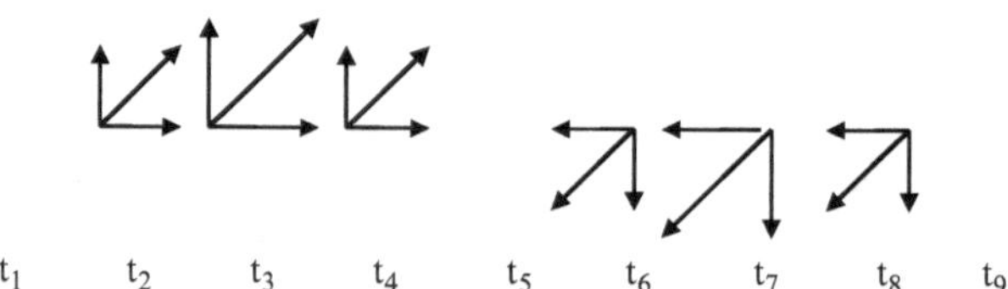

FIGURE 5.4 Total electric field when the field has both *up–down* components and *left–right* components; the components are equal, and they are "in phase". The figure assumes that we are looking along the direction of propagation of the photons, that is, the direction of propagation is into the page. The time intervals are spaced one eighth of a period apart. In each picture the diagonal arrow (*bolded*) represents the vector sum of the two field components, that is, it is the total electric field. At t_1, t_5, and t_9, both components (and therefore the total field) are zero.

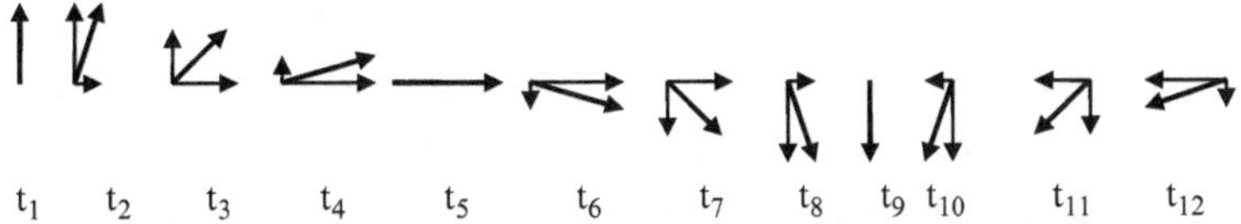

FIGURE 5.5 Total electric field when the field has both *up–down* components and *left–right* components, and they are out of phase. In each picture, the total field is bolded. As in Figure 5.3, the direction of propagation is into the page.

But the direction of the arrow is always in the plane that is 45° to both the vertical and horizontal axes. By the way, the amplitude of the total electric field, that is, its maximum value, increased a bit, and its vector got a bit longer, from the vector addition.

Now suppose that we added two electric field vectors so that the up–down vector was maximum up when the left–right one was zero. This situation is referred to as the vectors being "out of phase." In this case, the electric field vector appears to rotate, as indicated in Figure 5.5. However, if the maximum values of the up–down and left–right vectors are the same, the length of the total electric field vector will be a constant. Again, we are looking in the direction of propagation of the light. By changing the relative timing of the up–down and left–right components, we can make the direction of rotation go in the other direction. The one of these that appears to have the electric field direction rotating clockwise is referred to as right-circularly polarized light, while the one that rotates counterclockwise is left-circularly polarized light. As you might guess, scientists have invented fancier names for these to impress the nonexperts, but we don't need to worry about those names.

5.3 Relating Circularly Polarized Light and Molecular Chirality

As we have mentioned, the remarkable thing about circularly polarized light and the molecular chirality is that they are correlated. Left-circularly polarized light and right-circularly polarized light don't act the same way when they go through a medium containing molecules that have a selected chirality. This might take the form of attenuation of one form and transmission of the other, or it might be that the electric field direction gets rotated in different ways by the two molecular forms. Again, we don't need to worry about the details; suffice it to say that chirally selected molecules do impose different effects on the two forms of circularly polarized light, and that these effects are related to the structure of the molecules.

We know that all the amino acids that are naturally occurring on Earth, except one, are left-handed (and the exception is non-chiral). That is truly remarkable! It's also a problem for the Miller–Urey interpretation of the origin of life, since their experiment produced equal numbers of left- and right-handed amino acids. So how did the amino acids all become left-handed? This is a puzzle that has befuddled scientists for more than half a century. This has produced a variety of proposed solutions. We will discuss several of these in the following chapter. We will also discuss a newly minted suggested solution that circumvents many of the problems with the other solutions.

5.3.1 Chiral Selection

To understand how chiral selection of molecules might occur, we first have to ask how the different chiral forms of molecules might have become "separated," either in energy, or in space-time, or in some other way. To explore this we have to go back to Michael Faraday [7] who, in 1846, noted that molecules in an external magnetic field will tend to act as if they had an induced current of their electrons. This results from the fact that electrons in a magnetic field feel a force that is perpendicular to the field. Thus they will tend to move in circles, with the plane of the circles perpendicular to the direction of the magnetic field. Although this would not of itself produce a different effect on left- and right-handed molecules, it does set up this possibility. This effective rotating electron ring will produce something called a magnetic moment, which represents the interaction of the molecules with the external magnetic field.

However, particles and nuclei can also have their own intrinsic magnetic moments. In particular, if a particle, or a nucleus, has an intrinsic "spin", or degree of freedom that acts as if the particle or nucleus is also a rotating ring of charge (it isn't; it only acts that way), it will have such a magnetic moment. The vector that represents the magnetic moment points in the direction perpendicular to the plane in which the charge rotates. This is not as spooky as it might sound; think of the Earth rotating about its axis. Its vector representing its rotation points along the axis about which it rotates, that is, it is perpendicular to the planes in which things are rotating. Nuclei like ^{12}C and ^{16}O have an intrinsic spin of zero if they are in their ground states, so they are not candidates for this effect. However, ^{14}N does have a nonzero spin, which will interact with an external magnetic field and with the induced electron ring. Thus ^{14}N is crucial to the model I'm describing; the effect it produces could not exist for either ^{12}C or ^{16}O. And ^{14}N is a constituent that is common to all of the amino acids, so whatever outcome its spin has on the origin of amino acid chirality is guaranteed.

This outcome was described by Buckingham [8] in a paper that was designed to describe the effects of nonzero spin nuclei on nuclear magnetic resonance studies, and was subsequently extended by Buckingham and Fischer [9] and Harris and Jameson [10]. However, the effects Buckingham described for molecules in the magnetic field of a laboratory magnet would also apply to molecules in outer space, if they could be subjected to an external magnetic field. Nature conveniently produces the required conditions to apply the Buckingham effect to amino acids. Just such a field is produced in many nascent neutron stars, or black holes, as they are forming when a core-collapse supernova occurs. The magnetic field associated with a neutron star close to its surface can be billions of times as strong as the strongest magnetic field that scientists have produced in their laboratories on Earth, at least for greater than fractions of a second, so we might expect that these fields can produce extraordinary effects. They are important, but they require one other ingredient to produce amino acid chirality, and as the supernovae collapse to neutron stars or black holes they also supply it in abundance. It is the neutrinos, mentioned in Chap. 1, that cool the neutron star or black hole as it is forming and these are emitted in copious numbers.

What Buckingham found is that the interaction of the nuclear magnetic moment with the magnetic moment associated with the electrons that the Faraday effect had produced would act differently in a left-handed molecule than it would in a right-handed molecule. Thus the strong external magnetic field, together with the non-zero-spin nucleus ^{14}N, will allow Nature to distinguish between left- and right-handed molecules. But how would this couple to the chirality of the molecules?

5.3.2 A Little Atomic Physics and Some Very Basic Quantum Mechanics

To describe this, we need to do a little atomic physics and consider the effects of the quantum mechanical nature of atoms. I'll spare you from all the quantum mechanical details and just use the results. This won't make you an expert in quantum mechanics, but you'll have just enough information to be slightly dangerous. By the way, it helps to have some background in physics to understand the following explanation but if you don't have that background and complete understanding escapes you, don't worry. You don't have to digest all of this to follow our story. So here we go.

An atom will have a magnetic moment and a total angular momentum of $J\hbar$, where J can be either an integer: 0, 1, 2, etc., or a half integer: 1/2, 3/2, 5/2, etc., and where $\hbar$ is one of the constants of physics ("Planck's constant" divided by 2π). The interaction of the magnetic moment of the atom with an external magnetic field will produce a separation in energy of its "magnetic substates" that can be observed by detecting the photons produced or absorbed in the transitions between the magnetic substates. The magnetic substates are characterized by a quantum number that ranges from the value of the total angular momentum to minus the total angular momentum in units of one, for example, if $J=2$, the magnetic substate quantum numbers will be 2, 1, 0, −1, and −2. If $J=3/2$, the magnetic substate quantum numbers will be 3/2, 1/2, −1/2, and −3/2. You can think of these as representing the different orientations of the vector J with respect to the direction of the magnetic field. However, there is another catch, that being that in quantum mechanics, the magnitude of the total angular momentum vector is greater than the magnitude of any of its substates. For the $J=3/2$ case, the total angular momentum is a bit larger than 3/2, the length of its maximum projection. Since the total angular momentum is greater than the length of any of its components,

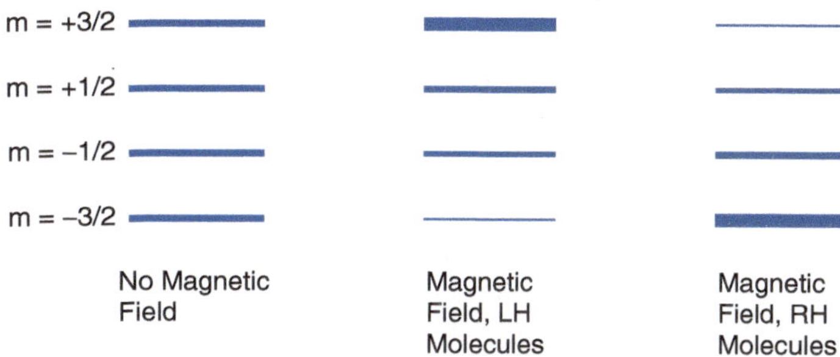

FIGURE 5.7 Populations of the different magnetic substates for an atom having angular momentum J=3/2 in a zero magnetic field (*left*), and for left-handed molecules (*center*) and right-handed molecules (*right*) in the presence of an external magnetic field. The thickness of the line indicates the magnitude of the population of each substate. (From Boyd et al. [11]. Courtesy of the International Journal of Molecular Sciences)

However, the Buckingham effect would redistribute the populations of these magnetic substates; with one chirality being driven toward one extreme and the other chirality toward the other. The situation is illustrated in Figure 5.7. On the left, the populations of the magnetic substates are seen to be equal, before the supernova goes off and the magnetic field is very small. But when the star explodes and the magnetic field forms, the left-handed molecules have their distributions driven one way, and the right-handed ones the other, as indicated in the figure.

How this distinction might manifest itself in a way that would produce a net chirality will be discussed in Chap. 7. But first we need to see, in Chap. 6, how some of the models for amino acid production have attempted to deal with the chirality problem. And we'll see in that chapter that chirality can indeed present a problem!

Are you beginning to feel a little bit dangerous?

5.4 The Other Molecules of Life

The very simple organic molecules, the amino acids, somehow can get together to form very complex molecules, the proteins. But one protein may have 100–1,000 of amino acids, and the 20

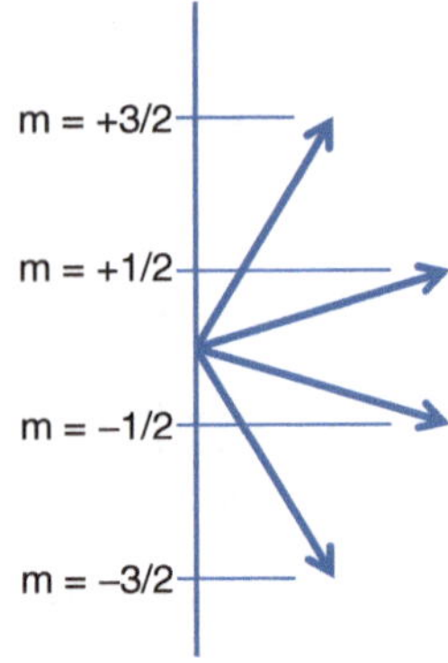

FIGURE 5.6 Orientations of the vector angular momentum J=3/2 for different orientations of J with respect to the external magnetic field. The magnetic substate quantum numbers m represent the possible projections of the total angular momentum on the "Z" axis, usually chosen to be the direction of the external magnetic field. Note that since the length of the angular momentum vector is larger than any of its projections, it can never be perfectly aligned with any of its components or with the external magnetic field

it can never be perfectly aligned with one of those components. Perhaps Figure 5.6 will help you visualize the situation.

The molecules with which we are concerned, the amino acids, will necessarily have many such total angular momenta, since different amino acids are composed of different atoms. For each amino acid, the total angular momentum will be the vector sum of the angular momenta of each of its constituent atoms. ^{14}N is not the only atom with a nonzero angular momentum: hydrogen atoms, for example, can also have a nonzero angular momentum, which will be the vector sum of the spin of its proton (1/2) and its electron (1/2). These can add to either zero or one. However, their magnetic moments will be small enough that the energy separations of their magnetic substates, even in the strong magnetic field that would be produced by a collapsing neutron star or black hole, would be small compared to the thermal energies of the atoms, even at the extremely low temperatures of outer space. Thus the thermal equilibrium that would exist would dictate that the numbers of molecules in each of the magnetic substates (that is, the populations) in any group of such atoms would be about the same, assuming that they have time to come to thermal equilibrium.

amino acids that are relevant to Earthly life could form gazillions of different proteins, only a small fraction of which would be relevant to life. So we are left to wonder how the amino acids get their instructions to form only the useful proteins.

Although there are a lot of questions associated with the origin, evolution, and operation of cells, this one does have an answer. As Davies says in his book The fifth Miracle, [12]"... life as we know it is the upshot of a mutually beneficial deal struck between DNA [deoxyribonucleic acid] and proteins." We'll talk a bit more about the DNA below, but for the moment, it contains the instructions for making the relevant proteins, down to the specific amino acid structure. DNA is constructed from four "nucleobases," adenine, guanine, cytosine, and thymine, abbreviated A, G, C, and T. We will also encounter another base, uracil, denoted by U, which is similar to T but which only occurs in RNA (see next paragraph). We will come back to these bases in a bit; their specific chemical character is crucial for replication, one of the basic criteria for achieving the status of a living being.

So the DNA has the instructions for forming the essential proteins, but it must communicate these instructions in some way so that the proteins get made. This is done via RNA, ribonucleic acid. RNA is made of AGCU, and comes in several varieties. Messenger RNA, or mRNA, reads the protein making instructions from the DNA and takes that information to the ribosomes, the places within the cell where the proteins are made. These have a "slot" into which the mRNA feeds the instructions selecting the amino acids and assembling them one by one until the protein is complete.

But while the assembly is taking place, somehow the right amino acid has to be identified and brought to the growing protein chain at the right time. This function is performed by another RNA, the transfer RNA, tRNA. The mRNA instructs the tRNA which amino acid it needs next, and the tRNA obligingly finds it, attaches to it, and delivers it to the site on the growing protein where it is needed. This process continues until the protein is completely formed, at which time the mRNA sends a signal to the ribosome, indicating that the process is complete and the newly created protein can be cut loose.

Of course, the DNA serves to do more than just providing the instructions for protein assembly. If you're not a biochemist, it's pretty easy to be blown away by descriptions of how DNA functions, but again Davies has given a nice description, which will form the basis and level for what follows.

In the 1950s, Francis Crick and James Watson, using state of the art technology (for that time) discovered that DNA has a structure of a double helix. The double strands are attached by many cross links, which are paired in such a way that A is linked with T and C with G. Thus the two strands are complementary, and when it comes time for the DNA to replicate, the two strands unwind, and each strand will combine with new bases to form (because of the complementarity of the two strands) a new strand that is identical to the one from which it just unwound.

One of the basic requirements for life is the ability to reproduce, and this clearly must persist for generations. It is this complementarity that produces the required unwavering replication of the DNA. The fact that human beings have been around for as many generations as we have provides testimony to the accuracy of the replication process. Of course, variations do occur, but these are rare. Furthermore, the DNA has mechanisms for correcting errors of replication. Of course, occasionally an error will remain uncorrected; these will form mutations. While most mutations are destructive, once in a while a positive one occurs, and this will provide its recipient with survival advantage, which gives the species with the new DNA a potential evolutionary gain.

5.5 How Were the More Complex Molecules Created?

The description of how the proteins are made and how the DNA replicates itself, of course, does not describe how everything got started. We do know that amino acids are produced in outer space, from the meteoritic evidence and from the space explorer Stardust, but how does Nature get from the amino acids to the more complex molecules, the proteins and the DNA? One model suggests that the peptides (formed from the amino acids) came first, and from them were formed the more complex molecules of life, the DNA and RNA; this is known as the "protein world." Peptides, which are smaller than proteins, generally contain less than 20 amino acids (proteins can contain many more) and have an amino acid on one end and a carboxyl on the other. The more complex molecules would then have formed initially as "oligomers," single stranded nucleic acid fragments. The other model suggests that somehow those more complex molecules had to be formed first, because without them how would the proteins get the instructions they required to form? This is called the "RNA world," which was originally proposed by Gilbert [13].

However, there is now experimental evidence that the amino acids can form peptides, and once that huge step has been taken, apparently the road is open for formation of more complex molecules. Surely the first ones were simple, but as mutations occurred, and natural selection oversaw the triumph of the more dominant forms, eventually the complexity that we now know was achieved.

The way in which the peptides are formed has been suggested to be through the Salt Induced Peptide Formation Reaction, or SIPF reaction [14]. Although theoretically based, the evidence for the reactions by which the peptides would be synthesized in a high concentration salt water environment are sufficiently well developed [15, 16] to warrant a detailed description of the process. The critical reactions were apparently catalyzed by copper ions in the salty environment [14]. The copper atoms served [17] "to bring [the amino acids] together into a suitable position and polarization state to allow peptide formation to take place with the help of the dehydrating agent NaCl" [16, 18]. These reactions, assuming a reaction set that is catalyzed by glycine (indicated as Gly), and X being another amino acid, are as follows. The dashes indicate bonds, and the process indicated by these equations forms a new chemical of the two X amino acids [14]:

$$\mathrm{Gly} + \mathrm{X} \rightarrow \mathrm{Gly} - \mathrm{X}$$
$$\mathrm{Gly} - \mathrm{X} + \mathrm{X} \rightarrow \mathrm{Gly} - \mathrm{X} - \mathrm{X}$$
$$\mathrm{Gly} - \mathrm{X} - \mathrm{X} \rightarrow \mathrm{Gly} + \mathrm{X} - \mathrm{X}$$

The copper organizer-catalyst may also serve another function. The effect on the energy difference between different chirality states due to the weak interaction goes as the fifth power of the nuclear charge which, with copper, is 29. Thus use of the copper catalyst has the potential to enhance enormously the energy difference due to the weak interaction between the two chirality states of molecules, and hence to produce a chirality selection based on this energy difference. Experiments have shown that, although the effect is small, it apparently does exist, and is reproducible [19].

One paragraph from the excellent review article by Rode, Fitz, and Jakschitz [17] deals with this question, and I will simply append that paragraph:

> For a final consideration of this problem—what was the first 'world' leading towards life?—one still has to address the question whether amino acids and peptides could have provided the necessary conditions which are required in the definition of life processes. Metabolism does not provide any difficulty in this aspect, whereas the questions of carrying on information, of self-replication, and of auto-catalytic processes have to be answered. While these processes are well-known to occur with RNA and DNA—although with the help of complicated protein-based enzymes—it has to be shown that these types of processes are also possible with peptides alone. In the eighties of the twentieth century, it was demonstrated that there are a number of peptides which are able to reproduce themselves from smaller constituents (Isaac and Chmielewski [20]; Lee et al. [21]; Yao et al. [22]), thus replicating their structure and their properties. On the other hand, it was found that amino acids themselves can catalyze peptide-formation processes, which can be seen as a good example of auto-catalysis in the production of oligomers (Suwannochot and Rode [23]). If one considers all of these aspects and findings, one arrives at the conclusion that the most likely way to initiate chemical evolution towards some primitive life forms was a beginning with a number of amino acids, forming peptides and later on the larger polymers, namely the proteins. Thus the question of RNA world or protein world does not seem any more a hen-and-egg problem but a clearly-to-answer problem in favor of a 'protein-first world,' starting from smaller peptides and forming in the end some cell-like organisms with first simple characteristics of life. One could ask at this point why one does not observe such primitive organisms any more. Here, one has to acknowledge that evolution has always replaced less efficient mechanisms by improved ones, eliminating thus the less favorable life forms. Hence, it is not difficult to postulate that as soon as RNA and DNA came into the 'game of life', any older organisms would have been eliminated.

However, an interesting twist has occurred, due once again to Stanley Miller. The famous Miller-Urey experiment showed that at least some amino acids could be formed in conditions and within environments thought to characterize the early Earth.

However, recent experiments have shown that a completely different environment is also capable of producing amino acids, and some nucleobases, the building blocks of DNA and RNA, as well. The first such experiment, which was performed by Miller, consisted of a solution he had prepared 25 years before he analyzed it. The experiment consisted of a dilute solution of ammonium cyanide, NH_4CN (which is equivalent to ammonia, NH_3 and cyanide, HCN), that was kept at two temperatures, –20°C and –78°C for 25 years (obviously this wasn't anyone's graduate thesis project). Although chemical reactions are generally thought to slow down as the temperature decreases, the reaction rates also depend on the densities of the entities that need to get together in order to react. As ice crystals form, they tend to drive out any impurities, which will then collect at the interfaces between the ice crystals, thereby vastly increasing the densities of the impurities. This is known as "eutectic freezing." To quote Fox [24] "Chemically speaking, [eutectic freezing] transforms a tepid seventh-grade school dance into a raging molecular mosh pit." The results were published in 2000 (Levy et al. [25]). Levy et al. found a simple set of amino acids, as well as adenine and guanine, two of the nucleobases.

Lower temperatures have other advantages also. Cyanide evaporates more readily than water, so operating at elevated temperatures tends to eliminate that constituent from a liquid mix. In addition, the nucleobases are rather fragile molecules, so lowering the temperature helps preserve them, extending their lifetime [24] from days to centuries.

Although the results of Levy et al. [25] were greeted initially with skepticism, probably because of the general feeling that reaction rates drop precipitously with decreasing temperature, additional experiments tend to confirm their result. Trinks et al. [26] conducted an experiment in which they "seeded" artificial sea water that was spiked with RNA nucleobases adenine, cytosine, and guanine, and with an RNA "template," a single-strand chain of RNA, to guide the formation of new strands of RNA. Then they froze the mixture for a year. As a new strand of RNA assembles, it "adheres to the template like one half of a zipper to the other" [24]. What Trinks et al. [26] found was that long strands of RNA molecules—up to 400 nucleobases long—had been formed. The previous record had been about 40 nucleobases.

These results certainly give hope to finding life on places like Europa (a moon of Jupiter) which is generally thought to be covered by a several kilometer thick layer of ice. However, it might also support the idea that not only amino acids, but nucleobases and even strands of RNA, might be formed in the cold confines of outer space. So perhaps the RNA world isn't dead. But one does have to ask where the template strand of RNA came from.

Of course, just having homochiral amino acids and nucleobases doesn't get us to the point at which these chemicals evolve into living beings. We certainly do not yet understand the chemistry by which this happens. However, a viable site in which the myriad chemical reactions required for this to happen from the molecules of life that are delivered to Earth from outer space has been suggested [27]: this is pumice. It is very porous, which gives it an extremely high surface to volume ratio, and presents the large surface areas on which the constituents necessary to create complex molecules and replicate the basic molecules of life, the amino acids and the nucleobases can gather. Perhaps even more importantly, this gives pumice a very low density, which makes it capable of floating on the surface of water, at least until it saturates. This allows it to pile up on shorelines so as to make large chunks of this chemical cauldron. Furthermore, the chunks of pumice might continue to serve as chemical cauldrons even when they sink, especially if they happen to end up near hot ocean vents, to which we will return in Chap. 9.

However the molecules of life formed, and life evolved from them, somehow life *did* get started from the complex molecules that preceded it. Could there have been just a single life form that evolved in a variety of different ways to form the plethora of living things that now inhabit Earth? The model that biologists have posited for this traces back to the Last Universal Common Ancestor, LUCA (see [1] for a more extensive discussion). What biologists have found is that all living creatures have many biological features in common. For example, even though no actual LUCA creature has ever been unearthed by our paleobiologists, the commonality of the species that "she" originated suggest that she relied on her DNA to store her genetic information, and that she used the same 20 (presumably left-handed!) amino acids to manufacture a couple of 100 proteins that her cells used to perform their various functions. (She was most likely a very primitive creature, perhaps even a single cell, after all; she might not be something that would be readily recognizable, especially after having been a fossil for a few billion years.) And LUCA may not have been unique; many other LUCAs may have been created at about the same time she was, but she was just the most successful in the competition for survival that had to have ensued.

References

1. K.W. Plaxco and M. Gross, *Astrobiology*, Johns Hopkins University Press, Baltimore, 2006
2. S.L. Miller, The Production of Amino Acids Under Possible Primitive Earth Conditions, Science 117, 528 (1953)
3. S.L. Miller and H.C. Urey, Science 130, 245 (1959)
4. A.P. Johnson, H.J. Cleaves, J.P. Dworkin, D.P. Glavin, A. Lazcano, and J.L. Bada, The Miller Volcanic Spark Discharge Experiment, Science. 322, 404 (2008)
5. A.I. Oparin, The Origin of Life on Earth, Oliver and Boyd, London (1957), originally 1936 in Russian, translated by Ann Synge
6. J.B.S. Haldane, Pasteur and Cosmic Symmetry, Nature 185, 87 (1960)
7. M. Faraday, Experimental Researches in Electricity, Nineteenth Series, Phil. Trans. Roy. Soc. London 1 (1846)
8. A.D. Buckingham, Chirality in NMR Spectroscopy, Chemical Physics Letters 398, 1 (2004)
9. A.D. Buckingham and P. Fischer, Direct Chiral Discrimination in NMR Spectroscopy, Chem. Physics 324, 111 (2006)
10. R.A. Harris and C.J. Jameson, J. A Note on Chirality in NMR Spectroscopy, Chemical Physics 124, 096101 (2006)
11. R.N. Boyd, T. Kajino, and T. Onaka, Stardust, Supernovae, and the Chirality of the Amino Acids, Int. J. Mod. Sci. 12, 3432 (2011)
12. P. Davies, *The 5th Miracle: The Search for the Origin and the Meaning of Life*, Simon and Schuster, NY, NY, 1999
13. W. Gilbert, The RNA World, Nature 319, 618 (1986)
14. D. Fitz, H. Reiner, K. Plankensteiner, and B.M. Rode, Possible Origins of Biohomochirality, Current Chem. Biol. 1, 000 (2007).
15. R. Tauler and B.M. Rode, Reaction of Cu(II) with Glycine and Glycylglycine in Aqueous Solutions at High Concentration of Sodium Chloride, Inorg. Chimica Acta 173, 93 (1990)

16. B.M. Rode and M.G. Schwendinger, Copper-Catalyzed Amino Acid Condensation in Water : A Simple Possible Way of Prebiotic Peptide Formation, Orig. Life Evol. Biosphere 20, 401 (1990)
17. B.M. Rode, D. Fitz, and T. Jakschitz, The First Steps of Chemical Evolution Towards the Origin of Life, Chemistry and Biodiversity 4, 2674 (2007)
18. M.G. Schwendinger and B.M. Rode, Possible Role of Copper and Sodium Chloride in Prebiotic Evolution of Peptides, Anal. Sci. 5, 1377 (1989)
19. K. Plankensteiner, A. Righi, B.M. Rode, R. Gargallo, J. Jaumot, and R. Tauler, Indications towards a Stereoselectivity of the Salt-Induced Peptide Formation Reaction. Inorg. Chim. Acta 357, 649 (2004)
20. R. Isaac, J. Chmielewski, Approaching Exponential Growth with a Self-Replicating Peptide, J. Am. Chem. Soc. 124, 6808 (2002)
21. D.H. Lee, J.R. Granja, J.A. Martinez, K. Severin, and M.R. Ghadiri, A Self-Replicating Peptide, Nature 382, 525 (1996)
22. S. Yao, I. Ghosh, R. Zutshi, and J. Chmielewski, Selective Amplification by Auto- and Cross-Catalysis in a Replicating Peptide System, Nature 396, 447 (1998)
23. Y. Suwannachot and B.M. Rode, Mutual Amino Acid Catalysis in Self-Induced Peptide Formation Supports This Mechanism's Role in Prebiotic Peptide Evolution, Orig. Life Evol. Biosphere 29, 463 (1999)
24. D. Fox, Did Life Evolve in Ice? Discover Magazine, Feb., 2008 Issue. Use of quotes courtesy of D. Fox
25. M. Levy, S.L. Miller, K. Brinton, and J.L. Bada, Prebiotic Synthesis of Adenine and amino acids Under Europa-Like Conditions, Icarus 145, 609 (2000)
26. H. Trinks, W. Schroder, and C.K. Biebricher, Ice and the Origin of Life, Orig. Life Evol. Biosphere 35, 429 (2005)
27. M.D. Brasier, R. Matthewman, S. McMahon, and D. Wacey, Pumice as a Remarkable Substrate for the Origin of Life, Astrobiol. 11, 725 (2011)

6. How Have Scientists Explained the Amino Acid Chirality?

Abstract In this chapter several of the models that purport to describe how amino acids are produced are discussed. Specifically, the means by which circularly polarized light could create chiral amino acids, both on Earth and in outer space, are detailed. Next, several models that rely on the weak interaction to select chirality are described. And, finally, the possibility that chiral selection might have occurred on solid surfaces is mentioned. In each case, the experimental documentation for the hypothetical model is presented. The chapter then discusses amplification mechanisms, notably, autocatalysis, which could operate either in outer space or on Earth. Finally, experiments that have demonstrated amplification toward homochirality are discussed.

6.1 Introduction to Models

Explaining amino acid chirality falls into two categories, and they are not necessarily distinct. It appears that no mechanism by which amino acid chirality is produced can generate more than a tiny amount of enantiomerism, that is, an imbalance toward one chirality. So any successful model requires that some mechanism or mechanisms exist that can produce some enantiomerism, and that at least one more mechanism must also exist that can amplify that enantiomerism to homochirality—the existence of a single chiral form.

There have been many explanations of the origin of amino acid chirality. For those astrobiologists whose models I don't discuss, but who think their model is superior to the ones I discuss, I apologize for overlooking your model, and invite an offline

R.N. Boyd, *Stardust, Supernovae and the Molecules of Life: Might We All Be Aliens?*, Astronomers' Universe, DOI 10.1007/978-1-4614-1332-5_6,

discussion. Perhaps your model will make it into the second edition! But I'll try to summarize some of them, especially the ones that I think are coming fairly close to being feasible explanations. Of course they fall into two classes, those claiming that the amino acids have a terrestrial origin, and those that claim they originated in outer space. And those in the former category still have to account for the chiral amino acids found in the meteorites.

6.2 Models That Produce Chirality

The suggestion that the amino acids were created by a lightning bolt in an appropriate Earthly environment was certainly given support by the Miller–Urey [1, 2] experiment, discussed in several previous chapters. This scenario, the Earthly creation of the amino acids, was reviewed extensively by Wachterheuser [3, 4]. Of course, the chirality issue presents a problem that this scenario must address. Thus many attempts have been made to explain how the equal populations of left- and right-handed amino acids that would have existed initially could have been converted to one that is entirely left-handed.

6.2.1 Chiral Selection via Circularly Polarized Sunlight

One possibility that has been suggested to make the stomach Miller-Urey model consistent with the Earth's amino acid chirality is to assume that our Sun's circularly polarized light might create some amount of enantiomerism as this light impinges on the amino acids, since one chirality of a molecule would then be more readily destroyed than the other (see, for example [5, 6]). This could happen under the following circumstances. A tiny fraction of the light from the Sun is polarized (<1%). It is usually assumed that the light has to scatter twice in order to achieve circular polarization, although we will discuss a model below in which that is not the case. We will assume that circularly polarized light is created when the light scatters from dust grains. The dust grains have to be non-spherical, in which case the interstellar magnetic fields (which are tiny but non-zero) can orient them. Then it can be shown that a single scattering will produce linearly polarized light (see Chap. 4). However, the orientation of the interstellar magnetic fields is pretty random, so if the light scatters a second time it will most likely be from a grain, again assumed to be non-spherical, that is oriented differently from the grain on which the first scatter occurred. If this is the case, the second scattering will change the phase of one of the components of the linearly polarized light, converting the linearly polarized light into circularly, or perhaps elliptically, polarized light. Since the ability of circularly polarized light to produce enantiomerism has been demonstrated in the laboratory, it should also work in nature.

The problem with the above model is the randomness of the magnetic field directions in interstellar space. Thus the circularly polarized light produced by the two scatterings is just as likely to create right-circularly-polarized light as it is to create left-circularly-polarized light. If the circularly polarized light creates left-handed molecules in one place, it will be circularly polarized in the opposite direction somewhere else, and so will make right-handed molecules there. The fractions of the two chiralities averaged over all space will be expected to be equal. Sunlight has an additional interesting feature: its circular polarization varies during the day. The different components during the day will just cancel each other. The attempts to explain how this "local" chirality but global equality of right- and left-handed molecules can be circumvented for Sunlight border on the heroic; they are discussed in the review paper by Bonner [7]. After thoroughly discussing them, he professes skepticism about their probability of success. Bailey [8] comes to the same conclusion.

The other possibility is that the amino acids were created in the cosmos, then transported to Earth by meteoroids, comets, or members of an advanced civilization. Front and center in this aspect of the discussion is the Panspermia hypothesis: that life was not created on Earth, but came to Earth from outer space. This was mentioned as early as the fifth century B.C., by the Greek philosopher Anaxagoras, and it has had many re-emergences since that time. Its proponents include Kelvin, von Helmholtz, Arrhenius, and in more recent times, Fred Hoyle, Chandra Wickramasinghe, and Francis Crick. Many of the Panspermia explanations have little in common with the original one except that they all have the origins of life external to Earth. Indeed, the purest version of Panspermia purports that life, and hence chirality, has always existed;

this would certainly force a revision of the modern theory of the Big Bang, since it is difficult to imagine molecules at the temperatures that existed in the early Universe, certainly the hundreds of millions Kelvin that existed at the time at which nucleosynthesis occurred a few minutes after the Big Bang, and even considerably higher than that prior to Big Bang Nucleosynthesis.

Nonetheless, the suggestion that the seeds of life did not originate on Earth but rather arrived in some way from outer space is certainly not new, and it was given a huge boost by the results of analyses of the Murchison meteorite, as we discussed previously. The Murchison meteorite showed that amino acids and other biologically interesting molecules are produced in outer space, that some of the amino acids have been chirally selected to be left-handed, and that they can survive the journey to Earth. Panspermia lives!

However, as often happens in science, this simply moves the question that needs to be answered to a higher, or in this case a more distant, level of sophistication. The nagging question of the origin of the chirality of the amino acids still remains to be answered and understood. The chiralities of the amino acids from the Murchison meteorite that were nonzero were all left-handed, and the amino acids on which we depend for life are left-handed. But is left-handedness inevitable for the amino acids? Is there some mechanism by which they are produced that always makes them left-handed? Or are they produced as racemic collections of molecules and then processed to make them enantiomeric? And can we use the chirality of the amino acids to understand the amino acid origins, and possibly even our origins? Indeed, is understanding the origin of the chirality of the amino acids crucial to understanding the origin of life?

6.2.2 Chiral Selection via Starlight

The possibility for producing left-handed amino acids via circularly polarized light can certainly be applied to making them in the interstellar medium as well as on Earth. Indeed, the photons may be more effective if they are higher energy than exists for the Solar photons, which must penetrate the Earth's atmosphere to get to the surface of the Earth. Thus ultraviolet photons, which cannot penetrate the Earth's atmosphere but are plentiful in outer space, may well be the preferred means of processing amino acids in space. Indeed, it has been demonstrated in the laboratory that this effect can occur [6]. It should also be noted that the Earth's atmosphere developed after the Earth was formed, and well after life is thought to have begun; thus the ultraviolet radiation could have had an effect on processing the amino acids that existed on Earth early in its history (although the Earth would still have had some sort of atmosphere, possibly the result of volcanic explosions and dust from meteorite impacts). The polarization appears to be highest when the photons have to pass through a relatively dense nebula in order to escape their star, which gives them more opportunities to scatter. However, starlight has been observed in most situations to be circularly polarized at only a fraction of a percent at the maximum and generally considerably less than that. So this scenario is roughly as promising for ultraviolet photons in outer space processing the amino acids as it is for sunlight affecting them on Earth. As with the model in which amino acids are processed on Earth by sunlight, the circularly polarized light would destroy molecules of one chirality and not affect the other, or more likely destroy both to some extent, but would affect one chirality more than the other.

Bailey [8] discusses the possible sources of circularly polarized light that might have resulted in amino acid chirality from ultraviolet photons in space. As a prelude to this discussion he presents a table that shows the amount of destruction of the amino acids that would be required to produce different enantiomeric excesses. If it is assumed that the light is 100% circularly polarized, destruction of 99.6% of the molecules would be required to produce an enantiomeric excess of 10%, and of 99.975% to produce 15%. If the light is only 1% circularly polarized, achieving an enantiomeric excess of 10% would require destruction of 99.996% of the molecules. This puts a high premium on performing the chirality selection with light that is as highly circularly polarized as possible.

The model developed by Gledhill and McCall [9] describes the polarization that would be realized from single scattering of ultraviolet light from non-spherical grains that would be aligned by a magnetic field. They found that grains having a ratio of longer dimension to shorter dimension of 2:1 could produce circular polarizations as high as 50%. And objects with circular polarizations approaching this value have been observed [10].

Bailey [8] discusses the possibility of various sources of circularly polarized light that have been discussed in the literature. This includes synchrotron radiation produced by relativistic electrons being accelerated in the magnetic field of a neutron star, which has not been observed to be polarized, so must have an extremely small polarization. It also includes the circularly polarized light from a white dwarf that is accreting matter from a companion star. This could produce an extremely high circular polarization, but Bailey argues that the probability that the material of which our Solar system is comprised would have been subjected to such a source is small. However, he also notes that if life in the universe is found to be rare, this might be a plausible source.

He finally concludes that reflection nebulae would seem to be the most promising source of the processed amino acids, especially given the model of Gledhill and McCall for producing the circularly polarized light. In contrast to the situation with the accreting white dwarf, Bailey notes that if life in the Universe is not rare, then this source would seem to be a major contributor. He also assumes that the processing would occur in a relatively dense cloud, so that the processed molecules would only undergo one processing event, and not even see photons from other stars.

However, this cosmic circularly polarized light scenario has the same difficulty producing global amino acid chirality as the circularly polarized Sunlight scenario, that is, the circularly polarized light that was left-handed in the cosmos would be expected to be balanced by the right-handed circularly polarized light. So this model might predict that there would be regions of the interstellar medium in which the amino acids would be left-handed, and others in which they would be right-handed, just as would be the case for the Sunlight scenario. And this assumes that the interstellar magnetic fields are not fluctuating, which might be expected to be the case for the starlight passing through the nebula produced by its star of origin, either from a stellar explosion or from stellar winds. But averaging over all space would make it difficult to produce anything other than a racemic result. This does remain a general problem for the circularly polarized light explanation in both the terrestrial origin and the extraterrestrial origin models as long as left-handed amino acids are all that we find. Another problem with circularly polarized ultraviolet light invoked to do the processing, as mentioned above, is also that the light will destroy a lot of molecules of life in order to create a preferred chirality in a few of them.

Another problem resides with the larger objects that might be processed by the circularly polarized light. The light would only process the surface of the object, so if it were a large meteoroid or a comet, the small fraction of the chirally selected molecules on the surface would translate into an infinitesimal enantiomeric excess when all the volume of the object, presumably with many more racemic molecules, was taken into account. And, since small objects would tend to be racemized or even destroyed when they encountered the Earth's atmosphere, it is presumably the large objects that delivered the Earth's preferred chirality. Bailey apparently circumvents this problem, however, by assuming that when the molecules were processed they would be either molecules in space or possibly could be on small dust grains. The dust grains would gradually clump together, and ultimately form much larger objects such as meteoroids. This would produce objects that would be uniformly chiral throughout, but would be large enough for a significant portion of each one to pass through the atmosphere of any planet on which they ultimately came to reside. However, timing is critical here, since the half-lives estimated for amino acids in a high photon flux environment are not long—possibly as short as several hundred years [11], so that grain formation must occur on a sufficiently rapid time scale to shield the molecules that had been selected. Once they are enclosed in a larger clump of material, their half-life increases to millions of years. However, grain clumping must not occur too rapidly, as the processing to establish chirality must occur on a shorter timescale.

An interesting approach described by deMarcellus et al. [12] assumes that the regions that are bathed in fairly uniformly circularly polarized light are large enough to encompass entire planetary systems. Technically we have very few data that tell us that the amino acids are all left-handed, and currently the data all originate within our Solar system (including those from comets). Thus it is possible that the star, or stellar region, that produced the circularly

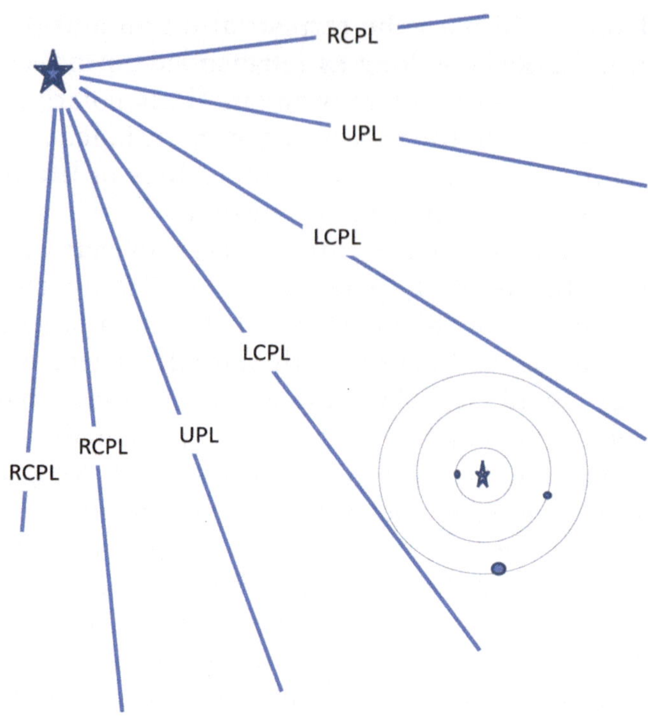

FIGURE 6.1 A sketch of a star that is emitting circularly polarized light (indicated by the outgoing rays), and a system of a star and planets (with orbits indicated) to which it appears that the circularly polarized light always has the same polarization. RCPL = right circularly polarized light, LCPL = left circularly polarized light, and UPL = unpolarized light. The average of the circular polarization over all space is zero, although large regions within a zone would make it appear that all the light was of one polarization. The radiation pattern shown is for illustrative purposes only; it is not intended to represent any known source of circularly polarized light

polarized light that made our Solar system with left-handed amino acids produced oppositely directed circularly polarized light that made another stellar system somewhere else right-handed. The global chirality would be zero, but we wouldn't know, yet anyway, about the right-handed planetary system. This model assumes that the processed regions did not get mixed, or there would be pockets of left-handed and pockets of right-handed amino acids. This situation is sketched in Figure 6.1.

However, should we discover a situation somewhere, for example, a comet or meteroid, in which right-handed and left-handed amino acids were equally probable in many of the inclusions of the comet or meteroid, or if astronomers were able to measure amino acids in some other planetary system that were right handed, this model would gain instant credibility.

6.2.3 Chiral Selection via the Weak Interaction

It might be thought that chirality could be achieved by a shift in the energetics of the molecules of opposite handedness, that is, if one of them might be selectively formed if it were more tightly bound than the other. In that case, thermal equilibrium would favor the more tightly bound molecule, which would then end up with the larger abundance. In several publications [13–21], Mason and Tranter studied the possible effects of the weak interaction (the one that mediates the beta decay we discussed in Chap. 3) in this regard. They found that the effect was extraordinarily small, about a part in 10^{17} (one part in 100 million billion). Gol'danskii and Kuz'min [22] conclude that this effect is so small that "an advantage factor stemming from weak neutral currents has not been solved on the basis of simple evolutionary [thermal] hypotheses." (Also note the discussion by Rode et al. [23] on the possible role of copper in peptide formation, and its possible role in establishing chirality via the enhancement it would produce on the energy separation of the chiral states.)

However, the weak interaction is the one known interaction that, because of its very nature, could produce a chirality because it "violates parity conservation." Parity conservation dictates that the mirror image of an object looks the same as the object itself, as we discussed in the introduction (remember your hands). This means that the weak interaction could produce a chirality simply as a result of the asymmetry that results from the interactions it might produce between particles. This property of the weak interaction was originally suggested by Lee and Yang [24], and confirmed by the Nobel prize winning experiment of Wu et al. [25]. This experiment showed that electrons emitted in the beta decay of ^{60}Co (Cobalt) nuclei that had their spins polarized in a strong magnetic field were preferentially emitted in the opposite direction to the spin, while the predictions of parity conservation would be that equal numbers would be emitted in the direction of the spin and in the opposite direction. The weak interaction is the

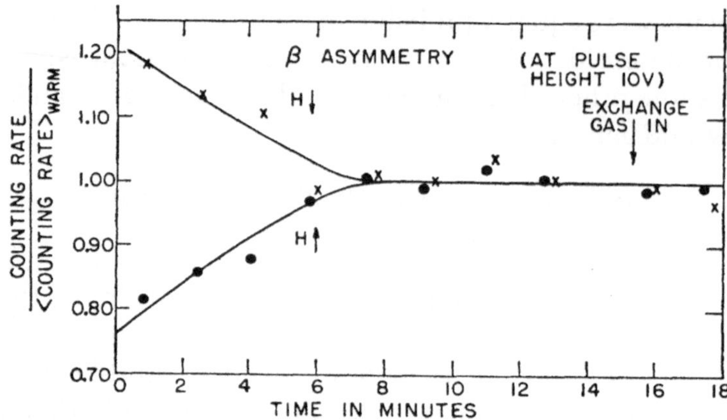

FIGURE 6.2 The data from the experiment of Wu et al. [25] to study the possible asymmetry of the decays from oriented ^{60}Co. The experiment is explained in the text. (From Wu et al. [25]. Copyright APS. http://link.aps.org/doi/10.1103/PhysRev.105.1413)

only one of the basic interactions in which parity is violated at a high level.

The results of the Wu et al. [25] experiment are shown in Figure 6.2. What is shown is the rate at which ^{60}Co nuclear decays were detected, in one case with the magnetic field (H) "up" and the other with it "down." The decay electrons could only be detected in the up direction because of the way the experiment was configured. The magnetic field provided orientation of the spins of the ^{60}Co nuclei, but only when the sample was cooled to near absolute zero. As time progressed and the sample warmed up, the orientation went to zero and the count rates with the two magnetic field orientations became the same. The data clearly show that there are more decays resulting from the field down configuration than from the field up configuration when the nuclei are oriented. If parity were conserved in beta decays, the rate at which electrons were emitted would not have depended on the nuclear orientation, and the event rates would not have depended on the magnetic field orientation. Thus it must be concluded that parity is not conserved in beta decay or in the weak interaction, which mediates beta decay.

Because this interaction violates parity conservation, it might be thought that it could produce an effect that could potentially skew the balance between left- and right-handed molecules. Indeed, since this is the only interaction that violates parity, it might be thought to be the obvious candidate to perform this function. One possible means by which this could come about originated with Vester et al. [26] and Ulbricht and Vester [27], who noted that the electrons produced in beta decay were longitudinally polarized (that is, the electron's spin is oriented antiparallel to its direction of motion) if they were sufficiently energetic. However, more energetic electrons tend to be more fully polarized. Highly relativistic electrons are essentially completely longitudinally polarized. In interacting with matter, these electrons would produce circularly polarized "Bremsstrahlung" photons, literally, braking radiation. This effect is produced when a charged particle (in this case an electron) is accelerated or decelerated when it encounters another charged particle. The Bremsstrahlung photons, in turn, could interact with the molecules to produce chiral selection.

The details of this scenario were studied by Mann and Primakov [28]. They attempted to show that the beta decay of ^{14}C, which despite being radioactive is a relatively abundant isotope of carbon because it is produced in the Earth's atmosphere, and which does undergo beta decay, could perform the chiral selection. They were able to show that this interaction could produce chiral selectivity, but only if there was a difference in the destruction rate of left- and right-handed molecules by the electrons, and that the expected effect was expected to be extremely small.

Several experiments [29–32] designed to demonstrate this effect produced results that suggested that some chiral selectivity did occur, but it was apparently due to the longitudinally polarized electrons, and not the Bremsstrahlung radiation they produced. It would seem that this would work as well as the Bremsstrahlung photons, but apparently the effects are still extremely small—barely at the level of detectability. Other attempts to confirm the existence of this effect have been frustrating, presumably because of the extremely low level of enantiomerism that this effect would produce [7, 33].

This effect was also studied theoretically by Gol'danskii and Kuz'min [22]. They concurred with Mann and Primakov [28] that the effect was small, but they actually gave a definitive estimate: it would be expected to be less than one part in 10^{10}. And this was an upper limit, since they assumed that the electrons were totally polarized. However, they also assumed the electron energy to be relatively low, where the probability for producing chirality is higher. But low energy electrons are less aligned than they would be at higher energies. The actual limit of the effect could easily be lower than was estimated by one or more orders of magnitude, and that would be consistent with the attempts to measure the size of the effect. Bonner [33] summarized the situation up to the time that article was written by concluding that parity violation in biomolecules was not likely to be the result of parity violations in the weak interaction.

However, recent work (as of 2011) using electrons from radioactive sources to bombard samples of chemicals such as CH_3OH, NH_3, and water, (H_2O) mixed with fairly large amounts of metals such as cobalt and copper have produced amino acids that were found to have tiny, but nonzero, enantiomeric excesses [34, 35]. Although the source of the enantiomerism produced in these experiments is not clear, this is an interesting result.

A related suggestion was made by Cline [36], who also advanced the idea of Vester and Ulbright. He noted that the incredibly intense neutrino flux from a core-collapse supernova might replace the electrons as the particles that might perform a chiral selection of the molecules with which they might interact. Cline focused on the number of reactions that the neutrinos, specifically, electron antineutrinos, would have with nuclei such as hydrogen. The relevant reaction is $\underline{\nu}_e + {}^1H \rightarrow e^+ + \text{neutron}$. No corresponding ("charged current") reaction can occur for electron neutrinos, since neutrons are unstable particles so are not present as isolated neutrons. Cline focused on the effects that the positrons, produced by the electron antineutrinos interacting with the protons, would have on the nuclei. However, he did not provide details of how this effect could produce molecular chirality. Cline also apparently did not consider what regions of space would be precluded from molecular interactions by the size of the progenitor of the core-collapse supernova, or by the photons produced in the supernova event. We will see in Chap. 7 that these are important to consider, and must be taken into account.

A similar suggestion [37] involved the interaction of supernova electron neutrinos and antineutrinos with atomic electrons. The interaction between the neutrinos and antineutrinos and the electrons would shift the energies associated with the different chirality molecules by a tiny amount, but by an amount that is comparable to the energies associated with thermal motion in the interstellar medium. However, this depends on the difference in abundance of the electron neutrinos and antineutrinos, which is not expected to be large. But if thermal equilibrium had enough time to prevail, and the energy difference was appreciable, this could (in principle) produce a preference for one molecular chirality over the other. However, as will be discussed in Chap. 7, the intense neutrino flux from a supernova lasts only a few seconds, and it is not clear that thermal equilibrium would have time to occur on that time scale in the cold environs of space. And this is even in addition to the requisite assumption of an appreciable difference in the numbers of electron neutrinos and antineutrinos.

6.2.4 Chiral Selection via Chemical Selection in Clay

A study by Fraser et al. [38] found that chiral selection in amino acids can occur in smectite clays such as vermiculite. These clays have expandable interlayer spacings, and the fresh layers exhibit significant chiral enrichment of either left- or right-handed amino acids, depending on the specific amino acid. The clays are certainly environments that could have existed in early Earthly environments, so this immediately becomes a potential way to perform chiral selection.

Of course it may be problematic for models of this type of chiral selectivity that some amino acids develop left-handed selectivity while others develop right-handed selectivity, unlike the apparent monolithic left-handed selectivity found in Nature. Furthermore, this model does not create or destroy one chirality, but rather separates them spatially. However, if either environment produced molecular replication more rapidly than the other, this would be a viable scenario for amino acid chirality selection. It is important to point out the possibility that other substances that might have existed in early Earthly environments could also produce this effect (such effects have been seen in quartz and calcite, as described in detail in Hazen [39]), and it would be important to sum over all such substances to see what the net effect would be. Most specifically, if other substances also selected different chiralities in different amino acids, and in some cases these tended to cancel the chiralities selected in clay, this could also be unfavorable to the clay model.

6.3 Amplification via Chemical Catalysis

The issues of both chirality production and amplification were studied in a seminal paper by Gol'danskii and Kuz'min [22]. As noted in Chap. 1, they observe that "chiral purity of the biosphere could not, even in principle, have been realized in the course of evolution, since without chiral purity of the medium the apparatus of self- replication could not occur." Gol'danskii and Kuz'min go on to show that the drive toward homochirality could occur either because one handedness achieved an "advantage," for example, one chirality might have become more prevalent than the other because of interactions with circularly polarized light (ignoring for the moment the difficulties with this scenario), or because the chemistry that creates new molecules from the required constituents might favor one chirality over the other.

In the latter context, they developed the model for amplification originally suggested by Frank [40] in 1953, and advanced by several astrobiologists since (see, for example, Kondepudi and Nelson [41]). As Frank originally conceived of the model, he was hypothesizing a substance that could act as a catalyst for its own production, and an anti-catalyst for production of the chiral opposite. In this model, left-handed molecule M_L and right-handed molecule M_R are both made from constituent molecules A and B. Once made, they can drive "autocatalysis," in which they can guide synthesis of new molecules of their same handedness from new molecules A and B. Finally, they can also combine to form a new molecule A′, destroying one M_L and one M_D in the process. For those of you who find it easier to understand these statements when they are written as equations, I've rewritten these three statements as the following equations where I've indicated to the right of each equation the rate constants, the k's, that dictate the speed at which each reaction occurs:

$$A+B \rightarrow M_L\left(k_1^L\right); \qquad A+B \rightarrow M_R\left(k_1^R\right)$$
$$A+B+M_L \leftrightarrow 2M_L\left(k_2^L\right); \quad A+B+M_R \leftrightarrow 2M_R\left(k_2^R\right)$$
$$M_L+M_R \rightarrow A'\left(k_S\right),$$

As noted by Gol'danskii and Kuz'min [22], Frank's original model did not have either of the reactions in the first line, nor did he include the possibility that the second set of reactions could go both ways.

What Gol'danskii and Kuz'min went on to show is that the relationship between the rates of these reactions can determine a critical value such that, if that critical value is exceeded, the succeeding reactions will drive the abundances toward a single chirality. With particular values of the reaction rates, the reactions involving molecules of one chirality might be formed over another, that is, one could have "asymmetric autocatalysis." They also note that this is "a consequence of the dynamic properties of the system itself and the process by which the mirror isomers (molecules of opposite chirality, but identical properties in all other respects, e.g., melting and boiling point, solubility, etc.) undergo transformations." Indeed, Gol'danskii and Kuz'min claim that these reactions could, in some cases, produce a single chirality even without an "advantage," that is, just from statistical fluctuations in the densities of the molecules.

However, an asymmetry in the reaction rates could also be important for amplifying existing enantiomerism via asymmetric autocatalysis, in which that asymmetry could favor reactions going toward one chirality over the other. This is an effect that has been demonstrated in a handful of laboratory experiments, although not necessarily for amino acids [42–44]. More generally, one could certainly enhance the drive toward homochirality by providing any advantage that would enable the dominance of one chirality over the other. Gol'danskii and Kuz'min divide the selective enhancement advantages into two classes, local and global. As the words imply, the local class would include an advantage that would occur only in a specific restricted location. This might include combinations of magnetic and electric fields, or circularly polarized light. Global advantages might include the effects of beta decay or some energetic shift between the left- and right-handed molecules, as discussed above, that always produce the same chirality.

6.4 Laboratory Experiments and Theoretical Developments

Several laboratory experiments have now been done that show how amino acids might be produced in cosmic dust grains. Bernstein et al. [45] showed that non-chirally selected amino acids (glycine, alanine, serine) could be produced in the lab via ultraviolet photolysis of interstellar ice analogs (H_2O, NH_3, CH_3OH, HCN). Chiral selection could then come later. Studies of this type have been going on for many years; a review article by Allamandola [46] gives some feeling for the scope of this work. Lee et al. [47] demonstrated that the amino acid glycine could be produced in an environment that simulated ice coated dust grains in the interstellar medium. This experiment radiated ice films that contained some of the basic molecules that have been observed in the cosmos, notably ammonia, NH_3, and methane, CH_4, with ultraviolet photons. Of course, the same ultraviolet radiation could destroy the molecules that had been produced. Thus another important result of the work of Lee et al. [47] was that it also demonstrated that some of the amino acids so formed would survive the ultraviolet radiation, particularly if the molecules existed in a dense part of a molecular cloud, so would be shielded to some extent from destruction. In any event, a balance between molecular production and molecular destruction would be established, assuming an equilibrium situation, so that some net amino acids would ultimately be created.

The ultraviolet radiation is not an essential feature of the cosmic organic molecule production. Bennett et al. [48] showed that energetic electrons, which simulate cosmic rays, and can be very abundant in the cosmos, especially in the vicinity of a star (think solar flares), could produce formic acid, HCOOH, by hitting mixtures of water and carbon monoxide ices.

Woon [49] investigated theoretically several "favorable reaction pathways" to production of some interesting organic molecules. He found that HCOOH, CH_3OH, and CO_2 could be created on simulated icy grain mantles from very low energy reactions, as would be expected in the cold cosmic environment, when the icy grain mantles interacted with several ions.

In a study of creation and amplification in the interstellar medium, Garrod et al. [50] developed a model in which chiral replication of complex molecules would occur in the warmed ice outer shells of grains. In their model, chemical replication would be catalyzed by radicals which are combinations of atoms that are not necessarily chemically stable (for example, H, OH, CO, CH_3, NH, and NH_2) created by the interactions of high energy cosmic rays with preexisting molecules. The grains would have to be warmed over the ambient temperature of the molecular clouds so as to enhance the mobility of the heavy radicals, but presumably this would occur periodically as the grains passed near a star and were subjected to the star's radiation.

6.5 Terrestrial Amplification

The ability of a collection of molecules exhibiting a small enantiomeric excess to amplify that excess dramatically has been demonstrated, and even in some environments that might plausibly exist on many planets. To give you a flavor of some of the experiments that have been done, I'll give a brief description of several of them that will not get too involved with organic chemistry! (1) Soai et al. [42] began with a left-handedness of 2% of 5-pyrimidyl alkanol treated with disopropylzinc and pyrimidine-5-carboxaldehyde. After performing an evaporation, they found that autocatalysis enabled by a chiral catalyst had increased the handedness to 85%. (2) Soai and Sato [43] began with a left-handedness of only 0.05% of methyl mandelate. Again, performing an evaporation, this was found to have been autocatalyzed to a much higher handedness. With the same procedure, Leucine (an amino acid) initially at a handedness of 2%, was enhanced to >95%. (3) Mathew et al. [51] began with an initial handedness of 5% of proline (an amino acid). After the evaporation this had a handedness of 65%. But they observed the interesting feature that the rate at which the reaction occurred accelerated as the chiral fraction increased.

Finally, Ronald Breslow and his student, Mindy Levine, [44] began with 1% chiral samples of right- or left-handed phenylalanine, which is also an amino acid. The samples were amplified to a chirality of 90% by two evaporations that precipitated out the non-chiral component. The conditions under which these experiments were conducted simulated conditions that could have plausibly existed in Earthly environments. Thus, although these experiments are interesting in their own right, they could have profound implications for the drive to homochirality of the amino acids, and of other life-related molecules. These experiments show that any model that can produce some preferred chirality, even at a low level, could have ultimately triggered a homochiral environment on its chosen planet.

Although the requirement of water in most or all of these experiments may seem to make these mechanisms unlikely in the interstellar medium, dust grains do develop icy surfaces, and a slight warming of these shells from the temperature characteristic of the interstellar medium could provide the conditions in which amplification could occur [47]. Thus the possibility that these processes could produce amplification in space should not be ignored. However, the water based experiments are certainly relevant for providing a second stage of amplification once the enantiomeric molecules arrived on Earth. And in several billion years of Earthly existence, a lot of meteorites might be expected to deliver a lot of enantiomeric amino acids to Earth. If most of them have the same chirality, this would surely enhance the drive toward homochirality.

6.6 Concluding Comments

I previously mentioned the model that my colleagues and I (Boyd et al. [52, 53]) developed; it also relies on the weak interaction, and on the neutrinos from supernovae. However, in addition to those commonalities with the Cline hypothesis [36], our model also requires the strong magnetic field produced by the supernova, and absolutely requires a non-spin-zero nucleus in order to couple to molecular chirality. Although this might be satisfied by the hydrogen in the amino acids, it turns out that the effect that they would produce is not nearly as definitive as that from ^{14}N, a constituent of all amino acids. We have dubbed our model the Supernova Neutrino Amino Acid Processing model: the SNAAP model.

If the SNAPP model is to succeed, or for that matter, if any of the suggested scenarios are to succeed, a lot of "amplification" is essential. In the SNAAP model, and probably in the circularly polarized light model, the chirally selected molecules need to replicate in outer space. As discussed above, it has been suggested that this might occur on the warmed surfaces of dust grains that are known to exist in outer space [50]. However, once the chirally selected molecules are bound up in a meteoroid and delivered to Earth in a meteorite, they need to be amplified again.

It appears that, for all of the models involving amino acid production and chiral selection in outer space, either autocatalysis or chemical replication via radicals could produce the amplification required to produce some initial chiral selection, and would maintain the chirality established by the magnetic fields and neutrinos from supernovae in the SNAAP model, or by the circularly polarized light in that model. The evaporation experiments discussed above could readily produce the additional amplification once the molecules were in Earthly environments that would ultimately produce homochiral molecule sets. But it appears that the SNAAP model, in which the chirality is selected by the supernova magnetic fields and the neutrinos, has the least negative caveats, and therefore presents the most robust scenario for the production of the initial enantiomerism of amino acids at a sufficiently high level that would lead to this ultimate homochirality. The means by which this might occur are discussed in the next Chapter.

References

1. S.L. Miller, The Production of Amino Acids Under Possible Primitive Earth Conditions, Science 117, 528 (1953)
2. S.L. Miller and H.C. Urey, Science 130, 245 (1959)
3. G. Wachterhauser, Before Enzymes and Templates: Theory of Surface Metabolism, Microbiol. Rev. 52, 452 (1988)

4. G. Wachterhauser, Groundworks for an Evolutionary Biochemistry: the Iron–Sulphur World, Prog. Biophys. Mol. Biol. 58, 85 (1992)
5. J. Bailey, A. Chrysostomou, J.H. Hough, T.M. Gledhill, A. McCall, S. Clark, F. Menard, and M. Tamura, Circular Polarization in Star-Formation Regions: Implications for Biomolecular Homochirality, Science 281, 672 (1998)
6. Y. Takano, J.-I. Takahashi, T. Kaneko, K. Marumo, and K. Kobayashi, Asymmetric Synthesis of Amino Acid Precursors in Interstellar Complex Organics by Circularly Polarized Light, Earth and Planetary Science Letters 254, 106 (2007)
7. W. Bonner, *The Origin and Amplification of Biomolecular Chirality*, Orig. Life Evol. Biosphere 21, 59 (1991)
8. J. Bailey, Astronomical Sources of Circularly Polarized Light and the Origin of Homochirality, Orig. Life Evol. Biosphere 31, 167 (2001)
9. T.M. Gledhill and A. McCall, Circular polarization by scattering from spheroidal dust grains., Mon. Not. R. Astron. Soc. 314, 123 (2000)
10. P.J. Hakala, V. Piirola, O. Vilhu, J.P. Osborne, and D.C. Hannikainen, Record Circular Polarization Discovered in the Shortest Period Magnetic Cataclysmic Variable RE 1307+535, Mon. Not. R. Astron. Soc. 271, L41 (1994)
11. P. Ehrenfreund, M.P. Bernstein, J.P. Dworkin, S.A. Sandford, and L.J. Allamandola, The Photostability of Amino Acids in Space, Astrophys. J. 550, L95 (2001)
12. P. deMarcellus, C. Meinert, M. Nuevo, J.-J. Filippi, G. Danger, D. Deboffle, L. Nahon, L.L.S. d'Hendecourt, and U.J. Meierhenrich, Non-racemic amino acide production by ultraviolet irradiation of achiral interstellar ice analogs with circularly polarized light. Astrophys. J. 727, L1 (2011).
13. S.F. Mason and G.E. Tranter, The Electroweak Origin of Biomolecular Handedness, Proc. R. Soc. London A397, 45 (1985)
14. S.F. Mason and G.E. Tranter, Energy Inequivalence of Peptide Enantiomers from Parity Non-Conservation, J. Chem. Soc. Chem. Comm. 117 (1983)
15. S.F. Mason and G.E. Tranter, The Parity-Violating Energy Difference Between Enantiomeric Molecules. Molec. Phys. 53, 1091 (1984)
16. S.F. Mason, Origins of Biomolecular Handedness, Nature 311, 19 (1984)
17. G.E. Tranter, Parity Violating Energy Differences of Chiral Molecules and the Origin of Biomolecular Chirality, Nature 318, 172 (1985)
18. G.E. Tranter, The Parity Violating Energy Difference Between Enantiomeric Reactions, Chem. Phys. Lett. 115, 286 (1985)

19. G.E. Tranter, The Parity Violating Energy Difference Between the Enantiomers of α-Amino Acids, Chem. Phys. Lett. 120, 93 (1985)
20. G.E. Tranter, Parity Violating Energy Differences and the Origin of Biomolecular Chirality, J. Theor. Biol. 119, 467 (1986)
21. G.E. Tranter, The Enantio-Preferential Stabilization of D-Ribose from Parity Violation, Chem. Phys. Lett. 135, 279 (1987)
22. V.I. Gol'danskii and V.V. Kuz'min, Spontaneous Breaking of Mirror Symmetry in Nature and the Origin of Life, Sov. Phys. Usp. 32, 1 (1989). Use of quotes courtesy of the American Institute of Physics, DOI: 10.1070/DU1989v032n01ABEH002674
23. B.M. Rode, D. Fitz, and T. Jakschitz, The First Steps of Chemical Evolution Towards the Origin of Life, Chemistry and Biodiversity 4, 2674 (2007)
24. T.D. Lee and C.N. Yang, Question of Parity Conservation in Weak Interactions, Phys. Rev. 104, 254 (1956)
25. C.S. Wu, E. Ambler, R.W. Hayward, D.D. Hoppes, and R.P. Hudson, Experimental Test of Parity Conservation in Beta Decay, Phys. Rev. 105, 1413 (1957)
26. F. Vester, T.L.V. Ulbright, and H. Krauch, Optical Activity and Parity Violation in beta-decay, Naturwissenschaften 46, 68 (1959)
27. T.L.V. Ulbright and F. Vester, Attempts to Induce Optical Activity with Polarized β-radiation, Tetrahedron 18, 629 (1962)
28. A.K. Mann and H. Primikov, Chirality of electrons from beta-decay and the left-handed asymmetry of proteins, Origins of Life 11, 255 (1981)
29. M. Akaboshi, M. Noda, K. Kawai, H. Maki, and K. Kawamoto, Asymmetrical Radical Formation in D- and L-Alanine Irradiated with Yttrium-90 β-Rays, Orig. Life Evol. Biospheres 9, 181 (1978)
30. M. Akaboshi, M. Noda, K. Kawai, H. Maki, Y. Ito, and K. Kawamoto, An Approach to the Mechanism of the Asymmetrical Radical Formation in Yttrium-90 β-Irradiated D- and L-Alanines, Orig. Life Evol. Biospheres 11, 23 (1981)
31. M. Akaboshi, M. Noda, K. Kawai, H. Maki, and K. Kawamoto, Asymmetrical Radical Formation in D- and L-Alanines Irradiated with Tritium β-rays, Orig. Life Evol. Biospheres 12, 395 (1982)
32. E. Conte, Investigation on the Chirality of Electrons from 90Sr-90Y beta-decay and their Asymmetrical Interactions with D- and L-Alanines, Nuovo Cimento Letters 44, 641 (1985)
33. W. Bonner, Parity Violation and the Evolution of Biomolecular Homochirality, Chirality 12, 114 (2000)
34. G.A. Gusev, K. Kobayashi, E.V. Moiseenko, N.G. Poluhina, T. Saito, T. Ye, V.A. Tsarev, J. Xu, Y. Huang, and G. Zhang, Results of the

second stage of the investigation of the radiation mechanism of chiral influence (RAMBAS-2 experiment). Orig. Life, Evol. Biosph. 38, 509 (2008)

35. V.I. Burkov, L.A. Goncharova, G.A. Gusev, K. Kobayashi, E.V. Moiseenko, N.B. Poluhina, T. Saito, V.A. Tsarev, J. Xu, and G. Zhang, First Results of the RAMBAS Experiment on Investigations of the Radiation Mechanism of Chiral Influence, Orig. Life Evol. Biosph. 38, 155 (2008)
36. D.B. Cline, Supernova Antineutrino Interactions Cause Chiral Symmetry Breaking and Possibly Homochiral Biomaterials for Life, Chirality 17, S234 (2005)
37. P. Barqueno and R. Perez de Tudela, The Role of Supernova Neutrinos on Molecular Homochirality, Orig. Life. Evol. Biosph. 37, 253 (2007)
38. D.G. Fraser, D. Fitz, T. Jakschitz, and B.M. Rode, Selective Absorption and Chiral Amplification of Amino Acids in Vermiculite Clay—Implications for the Origin of Biochirality, Phys. Chem. Chem. Phys. 13, 831 (2011)
39. R.M Hazen, Genesis: The Scientific Quest for Life's Origins, Joseph Henry Press, Washington DC (2005)
40. F. Frank, On Spontaneous Asymmetric Synthesis, Biochim. Biophys. Acta 11, 459 (1953)
41. D.K. Kondepudi and G.W. Nelson, Weak Neutral Currents and the Origin of Biomolecular Chirality, Nature 314, 438 (1985)
42. K. Soai, T. Shibata, H. Morioka, and K. Choji, Asymmetric Autocatalysis and Amplification of Enantiomeric Excess of a Chiral Molecule, Nature 378, 767 (1995)
43. K. Soai and I. Sato, Asymmetric Autocatalysis and its Application to Chiral Discrimination, Chirality 14, 548 (2002)
44. R. Breslow and M.S. Levine, Amplification of Enantiomeric Concentrations Under Credible Prebiotic Conditions, Proc. National Acad. Sciences 103, 12979 (2006)
45. M.P. Bernstein, J.P. Dworkin, S.A. Sandford, G.W. Cooper, and L.J. Allamandola, Racemic Amino Acids from the Ultraviolet Photolysis of of Interstellar Ice Analogues, Nature 416, 401 (2002)
46. L.J. Allamandola, Chemical Evolution in the Interstellar Medium; Feed Stock in the Solar Systems, in Chemical Evolution Across Space and Time – From the Big Bang to Prebiotic Chemistry, ed. By L. Zaikowski and J.M. Friedrich, 80 (2008)
47. D.H. Lee, J.R. Granja, J.A. Martinez, K. Severin, and M.R. Ghadiri, A Self-Replicating Peptide, Nature 382, 525 (1996)

48. C.J. Bennett, T. Hama, Y.S. Kim, M. Kawasaki, and R.I. Kaiser, Laboratory Studies on Interstellar and Cometary Ices, Astrophys. J. 727, 27.01 (2011)
49. D.E. Woon, Ion-Ice Astrochemistry: Barrierless Low-Energy Deposition Pathways to HCOOH, CH_3OH, and CO_2 on Icy Grain Mantles from Precursor Cations. Astrophys. J. 728, 44–1 (2011)
50. R.T. Garrod, S.L.W. Weaver, and E. Herbst, Complex Chemistry in Star-Forming Regions: An Expanded Gas-Grain Warm-up Chemical Model, Astrophys. J. 682, 283 (2008)
51. S.P. Mathew, H. Iwamura, and D.G. Blackmond, Amplification of Enantiomeric Excess in a Proline-Mediated Reaction, Angew. Chem. Int. Ed. 43, 3317 (2004)
52. R.N. Boyd, T. Kajino, and T. Onaka, Supernovae and the Chirality of the Amino Acids, Astrobiology 10, 561 (2010)
53. R.N. Boyd, T. Kajino, and T. Onaka, Stardust, Supernovae, and the Chirality of the Amino Acids, Int. J. Mod. Sci. 12, 3432 (2011)

7. What Happens to the Amino Acids When the Supernova Explodes?

Abstract This is the chapter in which we will put all the things we have been discussing together, along with a few new entities, to form the SNAAP model. This will necessarily involve several important aspects of supernova astrophysics, some neutrino physics, interactions of neutrinos with nuclei, some basic nuclear physics, some physics conservation laws, and finally, some properties of Wolf-Rayet stars and of red giants. Along the way, I'll try to give you some interesting facts about some of the entities we encounter, especially the neutrinos.

7.1 The Supernova

Let's start with the supernova. Just to remind you, what we are talking about here (as described in Chap. 3) are core-collapse supernovae. When a supernova occurs, the core of the star contracts into either a neutron star or a black hole. However, in the latter scenario, the supernova may first contract to a neutron star, then collapse to the black hole after some material that was initially blown outward falls back onto the neutron star, causing it to exceed the maximum mass that a neutron star can have. Like a white dwarf, the neutron star also has a maximum mass, although it's supported by the degeneracy pressure of the neutrons and the few protons that are there, instead of the degeneracy pressure of the electrons, as is the case for a white dwarf. (Remember, the degeneracy pressure is what maintains the size of atoms, where only one electron is allowed to occupy each quantum state.) The collapse to the black hole might take a little time, perhaps a

R.N. Boyd, *Stardust, Supernovae and the Molecules of Life: Might We All Be Aliens?*, Astronomers' Universe, DOI 10.1007/978-1-4614-1332-5_7,

few seconds or so, but that is plenty of time for a lot of neutrinos to be emitted. The general result is that, independent of the final state of the star, a lot of neutrinos will be emitted as the final contraction occurs.

What happens when the neutron star collapses to a black hole? This doesn't take very long, but the collapse affects things other than the neutrinos in different ways. The parts of the star that got exploded outward during the initial explosion may be able to remain outside the black hole. But the stuff that was infalling when the core object became a neutron star is not likely to escape. What happens to all the photons that were spewed out from the center of the star when it exploded? The photons scatter from all the matter that surrounds the core, and it takes them perhaps an hour to appear to the outside world. But that's a very long time compared to the collapse time of the core, so those photons are likely to be trapped by the infalling matter, and to disappear inside the event horizon of the black hole. These supernovae won't produce much of a show compared to those that don't collapse to black holes; indeed they would not be seen at all.

Supernovae that do explode are such gorgeous objects that I had to include a picture of one. It's shown in Figure 7.1 for "Kepler's supernova," first seen on Earth in 1604 AD. The figure shows the supernova remnant as it now appears to several of NASA's orbiting observatories. One, the CHANDRA Observatory, observes X-rays which are higher energy photons than could be seen in the optical, that is, visible, wavelengths. As you can see from Figure 7.1, this object is pretty dim at the optical wavelengths. As we discussed earlier, photons that are just a little bit more energetic than what our eyes can see are called ultraviolet light, and they cannot penetrate the Earth's atmosphere. The Hubble Space Telescope can observe ultraviolet light that is just beyond the visible part of the spectrum since it orbits above the Earth's atmosphere. The Spitzer Space Telescope was designed to see "infrared light"—lower energy photons than can be seen by our eyes. Neither the x-rays seen by CHANDRA nor the infrared photons seen by Spitzer can penetrate the earth's atmosphere, so orbiting telescopes are essential for viewing them. The larger picture puts the separate components together with the colors associated with the different

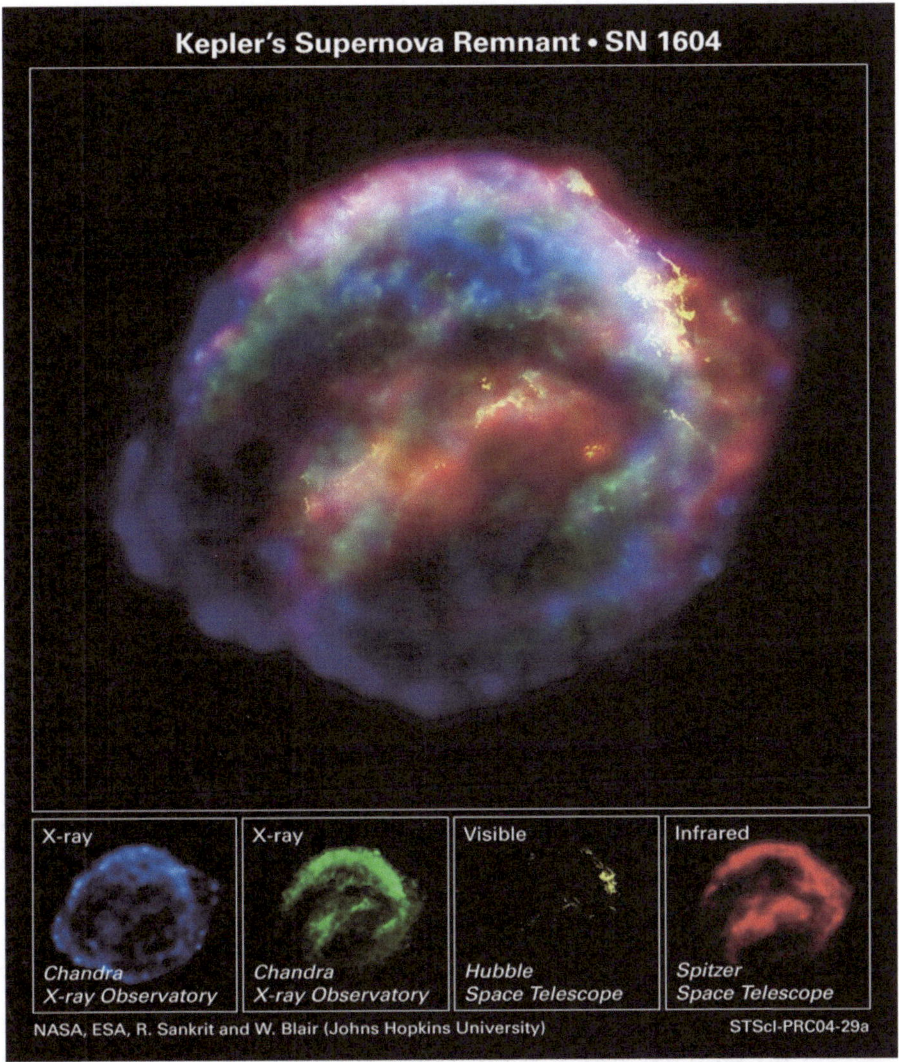

FIGURE 7.1 Kepler's supernova SN 1604. (Courtesy of Space Telescope Science Institute and NASA. Image by R. Sankrit and W. Blair)

observatories, as indicated below the figure. Since our eyes can't see x-rays, or infrared light, this isn't what the supernova would actually look like if you were observing it with a telescope. But I'm sure you would agree that this "reconstructed" supernova is gorgeous!

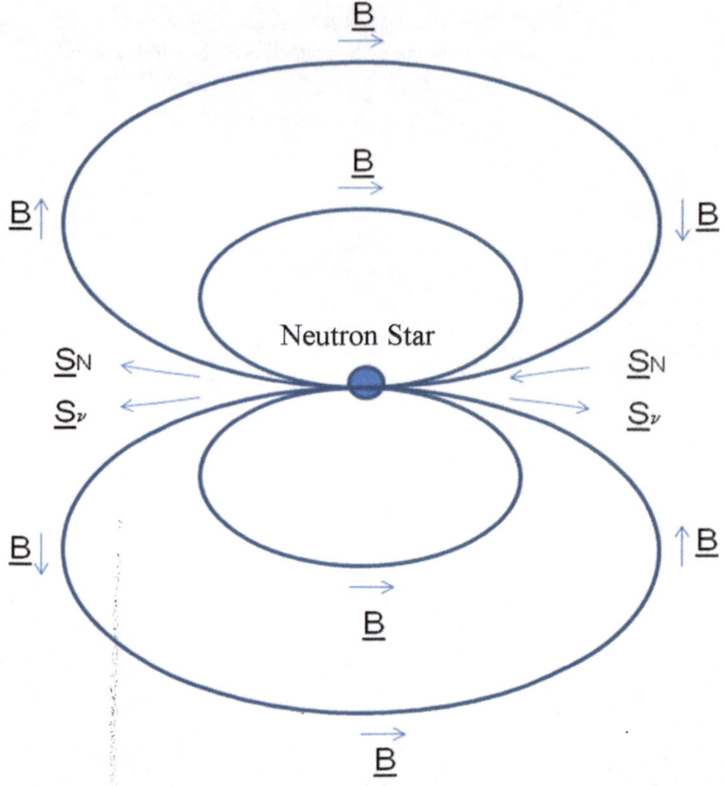

FIGURE 7.2 Magnetic field around a neutron star, indicated by B; directions of the antineutrino spins, indicated by $\underline{S}_v$, and the alignment direction of the ^{14}N, indicated by $\underline{S}_N$ from Boyd et al. [1, 2]. (Courtesy of Astrobiology, and of the International Journal of Molecular Science)

As previously noted, there is another aspect of this collapse that is important to our story. This is the magnetic field that is often generated in the collapse to the neutron star or black hole. The magnetic fields associated with these objects are prodigious at the surface of the neutron star; they can be billions of times as large as anything we can generate in our terrestrial laboratories. The magnetic field lines that are produced are shown in Figure 7.2; they spread out over space as indicated. We've shown these before, but now we are going to show how they can produce the interactions with the molecules that are needed to distinguish between molecules of different chirality.

The thing to note in Figure 7.2 is that at one pole of the neutron star the magnetic field is outgoing and at the other side it's incoming. This is because the magnetic field lines have to close on themselves, that is, they need to form closed loops. The magnetic field from the neutron star will interact with the magnetic moment of the ^{14}N (recall that is the property of particles that makes it look as if the nucleus has a current loop running around inside of it). And the magnetic moment is related to the spin—that intrinsic degree of freedom that makes it look as if the ^{14}N is rotating about its axis (recall that is analogous to the Earth's rotation about its axis). The spin acts as if it is an angular momentum, which for microscopic particles such as nuclei is measured in units of Planck's constant, that extremely small fundamental constant of quantum mechanics. Because the magnetic moment interacts with the neutron star's magnetic field, it will also be pointing outward on one side of the neutron star and inward on the other, as is shown in Figure 7.2.

7.2 The Neutrino Story

Now we need to discuss what the neutrinos are doing. But first I need to tell you some things about neutrinos. Some of this is important to our story, and some I'm just telling you because neutrinos are really cool particles, and if you're interested in them, I'll give you a bit of background. As noted in Chap. 1, neutrinos are chargeless, nearly massless particles that interact only through the weak interaction. Well, they also interact through the gravitational interaction, but that doesn't affect the present story, and is irrelevant for virtually every situation except those on very large scales like motions within galaxies and universes, just because the neutrinos have such tiny masses.

Neutrinos come in three types, or "flavors," electron neutrinos, mu neutrinos, and tau neutrinos, and each has its corresponding antiparticle. Electron neutrinos are produced when a nucleus has too many protons to be stable, as discussed in Chap. 3. This process is called "beta-decay," in which the nucleus will either capture an electron to convert one of its protons to a neutron or emit a positron, which is an anti-electron, again to convert one of its protons to a neutron. (You may recall that an anti-electron has the mass and spin of the electron, but with a positive instead of negative charge.) These two decay reactions look like this:

$$\text{Nuc}(Z,N)+e^{-} \rightarrow \text{Nuc}'(Z-1,N+1)+\nu_e$$

or

$$\text{Nuc}(Z,N) \rightarrow \text{Nuc}'(Z-1,N+1)+e^{+}+\nu_e,$$

In the above reaction, ν_e is the electron neutrino, e^+ is the positron, Nuc(Z,N) refers to a nucleus with Z protons and N neutrons, and Nuc′(Z–1, N+1) refers to a new nucleus with one less proton and one more neutron than existed in Nuc(Z,N). Both processes require that the mass-energy of Nuc(Z,N) be greater than that of Nuc′(Z–1, N+1)+the mass energy of the positron. As discussed in Chap. 3, mass-energy here refers to Einstein's famous equation $E=mc^2$ (physicists often express masses in terms of their mass-energies).

If that mass-energy difference is positive but small, only the first process will occur. If it is greater than about 1.5 MeV, both processes can occur, but the second one will dominate. "MeV" stands for Million electron Volts, that unit of energy that is especially useful for describing nuclei and nuclear processes, which are often characterized by energies that are the order of an MeV. As an example, ^{15}O is a radioactive nucleus which is produced in the hydrogen burning in massive stars, as discussed in Chap. 3. It has eight protons and seven neutrons, and decays to ^{15}N (which has seven protons and eight neutrons), a positron, and an electron neutrino. The mass-energy of ^{15}O is greater than that of ^{15}N+ that of the positron by 2.754 MeV.

Conversely, if a nucleus is formed that has too many neutrons to be stable, it will emit an electron and an electron antineutrino. This is also called beta-decay. This occurs as a critical part of either the s-process or the r-process of nucleosynthesis, both of which occur in stars to make heavy nuclei, and which were discussed in Chap. 3. This beta-decay process looks as indicated below, where e^- is an electron and ν_e denotes the electron antineutrino:

$$\text{Nuc}(Z,N) \rightarrow \text{Nuc}''(Z+1,N-1)+e^{-}+\underline{\nu}_e,$$

In both forms of beta-decay, a "beta-particle," that is, an electron or a positron, is involved either being emitted or captured. An electron antineutrino or neutrino is emitted in both cases.

At the sorts of temperatures, hence energies, that exist in the Sun, only electron neutrinos and electron antineutrinos could be produced, but actually, only electron neutrinos are emitted. That's due to the nuclear reactions that take place in the Sun, as explained in Chap. 3.

Neutrinos and antineutrinos are all spin ½ particles, but the neutrinos and the antineutrinos have opposite chirality. The ones about which we will be concerned are the electron neutrinos and antineutrinos; the spin of the electron antineutrino will be parallel to its direction of motion, whereas that of the electron neutrino will be antiparallel to its direction of motion. This will play a crucial role in the selection of molecular chirality, as we will see below.

However, at higher energies, for example, those that take place when high energy cosmic rays hit the Earth's atmosphere, "pions" can be produced. You must be thinking that this zoo of particles is never-ending, but we are almost there, at least for this book. If the book were about particle physics, or if I had included a discussion of ultra-high-energy cosmic rays, we'd just be getting started. Pions are unstable particles that can be produced when the energies of the colliding particles exceed the value of the pion mass-energy, $m_\pi c^2$. For physics aficionados, (and you might be one by now, having gotten this far) pions are composed of a quark and an antiquark. The pion quickly decays into a muon and the appropriate mu-neutrino, as indicated below. The symbols π^+ or π^- denote a positively or negatively charged pion (there can also be a pion with zero charge, but it decays quite differently), μ^+ or μ^- denote a positively or negatively charged muon, and ν_μ and $\underline{\nu}_\mu$ denote, respectively, a muon neutrino and a muon antineutrino.

$$\pi^+ \rightarrow \mu^+ + \nu_\mu$$

or

$$\pi^- \rightarrow \mu^- + \underline{\nu}_\mu,$$

Just in case you were wondering, when the muon decays, it decays into an electron if it's a μ^-, or a positron if it's a μ^+, and two more neutrinos. These decays look like:

$$\mu^+ \rightarrow e^+ + \underline{\nu}_\mu + \nu_e$$

or

$$\mu^- \rightarrow e^- + \nu_\mu + \underline{\nu}_e.$$

There is another physics conservation law that applies to these decays, that is, "lepton conservation." In each reaction the number of leptons, (the electrons, muons, neutrinos, and their antiparticles), on the left hand side of the equation must be the same as their number on the right hand side. A lepton counts as +1 particle, and an antilepton as −1. So, for example, in the last equation, the μ^- on the left hand side counts as +1 lepton, as does the e^- on the right hand side. The ν_μ counts as +1 lepton, while the $\underline{\nu}_e$ counts as −1. So everything adds up as it should.

To complete the story, tau particles, which are also leptons, can be produced in yet higher energy interactions. They have such a short lifetime that they barely move at all, unless they are highly relativistic, which will lengthen their decay time by time dilation, an effect that exists for particles moving near the speed of light (and also for space travelers moving near the speed of light). However, when they decay, they will produce tau neutrinos, just as the muons produced mu neutrinos.

7.3 Interactions of Neutrinos with ^{14}N

Returning now to the lower energies relevant to our story, we will concentrate on the electron antineutrinos and electron neutrinos that can change ^{14}N to either ^{14}C or to ^{14}O respectively. In both cases we're relying on "charged-current weak interactions" to mediate the process. That's a mouthful, but I include that to discriminate between that interaction and "neutral-current weak interactions," which would produce different effects that wouldn't be chirally selective. In addition, I should note that the cooling neutron star can emit not only electron neutrinos and antineutrinos, but mu- and tau- neutrinos and antineutrinos. Those neutrinos can be produced deep inside the star as neutrino-antineutrino pairs, a process that requires much less energy—only $2m_{\nu}c^2$, and recall that the neutrino masses are all really tiny—than if it were necessary to produce a muon or a tau particle via a charged-current weak interaction.

Although the energies of the neutrinos from a supernova can get to be pretty high, they don't become high enough to exceed the thresholds imposed by the requirement that the energies must exceed the rest mass energy of the muons–105 MeV in rest mass energy units. (The electron mass in those same units is 0.511 MeV or for taus is it 1777 MeV.). The electron neutrinos and antineutrinos operate on ^{14}N by the following reactions:

$$^{14}N + \underline{\nu}_e \rightarrow {}^{14}C + e^+$$
$$^{14}N + \nu_e \rightarrow {}^{14}O + e^-.$$

In the first reaction, the electron antineutrino $\underline{\nu}_e$ converts ^{14}N to ^{14}C, and in the second, the electron neutrino ν_e converts ^{14}N to ^{14}O. Note that these reactions look just like the beta decays that we discussed in Chap. 3 in the context of hydrogen burning in massive stars, except that here the neutrino appears on the left side of the equation instead of the right side. And indeed these reactions are mediated by the weak interaction, just as is beta decay. Also, if you switch a neutrino, antineutrino, electron, or positron from one side of the equation to the other, you need to change it to its antiparticle (to obey the lepton conservation law). The neutrino energies required to perform these reactions are important, as the reaction probabilities depend on the energy.

There is considerable controversy about what the actual neutrino energies from a supernova will be. However, the studies of Georg Raffelt and his group [3, 4], and of George Fuller and his group [5–7], suggest that so much mixing of the neutrino flavors, or types, will take place in the stars that all six of the neutrino and antineutrino flavors will emerge from the star with similar energy distributions, that is, their "energy spectra" will look very similar.

One prediction of the neutrino energies, from Gava et al. [8], is shown in Figure 7.3. Although it shows what the expected energies of the electron antineutrinos will be, it is roughly representative of

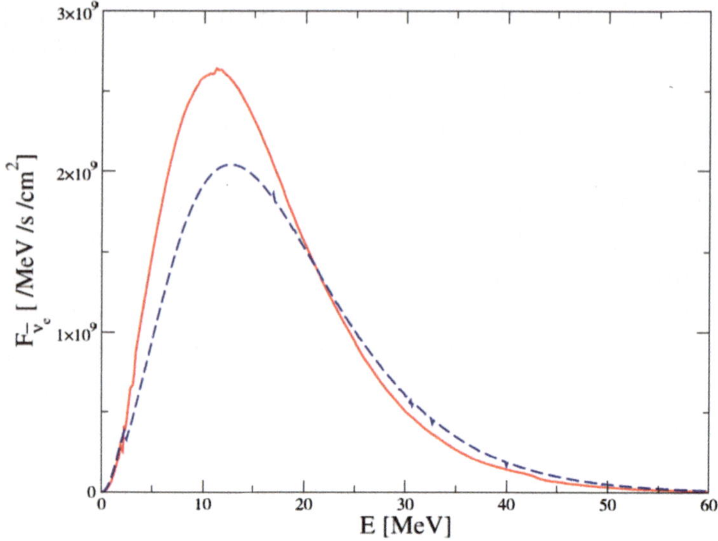

FIGURE 7.3 Energy spectrum from a core-collapse supernova that is typical of all six neutrino and antineutrino flavors [8]. The two curves represent two different assumptions of neutrino processing. (From Gava et al. [8]. Copyright APS. Courtesy of G.C. McLaughlin. http://link.aps.org/doi/10.1103/PhysRevLett.103.071101)

all the neutrino types. Note that the peak energy is around 12 MeV in either of the two neutrino interaction scenarios represented in that figure.

This energy is important because the "threshold energy" for the $^{14}N + \underline{\nu}_e \rightarrow {}^{14}C + e^+$ reaction is 1.18 MeV (that is the energy that the neutrino or antineutrino is required to have for the reaction to take place), while the threshold energy for the $^{14}N + \nu_e \rightarrow {}^{14}O + e^-$ reaction is 5.14 MeV. Thus both reactions are endothermic; they require energy for them to occur. That energy has to be supplied by the incident antineutrino or neutrino. This means that many more of the electron antineutrinos will be able to convert ^{14}N to ^{14}C than will the electron neutrinos be able to convert ^{14}N to ^{14}O. In addition, most models of the neutrinos emitted from supernovae have the electron neutrinos somewhat less energetic, by roughly 2 MeV, than the electron antineutrinos. Finally, the reaction probabilities are proportional to the square of the amount of energy the neutrino has in excess of the threshold energy; this will further enhance the importance of the reaction that converts ^{14}N to ^{14}C

compared to that which converts ^{14}N to ^{14}O. Suffice it to say that, given the expected energies of the neutrinos from a supernova, the conversion to ^{14}C is considerably more probable than the conversion to ^{14}O. So we'll ignore the conversion to ^{14}O for now, but come back to it a bit later.

Now we need to do a little nuclear physics. I'll give you the summary of what happens, just in case you don't want to get too involved in the details of the nuclear physics (I've heard that such people do exist!). But the details I'll present are pretty simple. Basically what happens is this: when the electron antineutrino and ^{14}N spins line up, this makes it more difficult for the antineutrino to do the conversion to ^{14}C than if they are not aligned. This will be related to the chirality of the molecules, as we will discuss below, so this will produce a chirality selective destruction at each throat of the neutron star. Although the effects at the two throats of the neutron star will tend to cancel each other, the neutrinos are not emitted in equal numbers at the two throats, which will result in a chirality selection.

The details involve more vector addition, but you mastered that in Chap. 5, so this will be easy. The spin of the neutrinos is 1/2 (in units of $\hbar$, of course), so when the spin of the antineutrino and the spin 1 ^{14}N nuclei are aligned, the total spin is 3/2. When they are antialigned it is 1/2. If you haven't studied your quantum mechanics, you might think it could be other values, for example, 5/8 or 2/3 or something like that, but it can't; the rules of quantum mechanics don't allow any but integer or half integer spins and angular momenta.

The laws of physics require that total angular momentum (spin is a kind of angular momentum) be conserved in any reaction, that is, the sum of the angular momentum vectors before an interaction or a decay must be the same as their sum afterward. So what happens is shown in Figure 7.4. The positron on the right hand side of the equation above has a spin of 1/2, so if the total spin was 1/2, as for the antialigned case, the total spin before the reaction is equal to the total spin after the reaction, since the final nucleus, ^{14}C, has zero spin (indicated by just the dot in Figure 7.4). But in the aligned case, that will not be the situation unless we come up with one additional unit of angular momentum from somewhere. That will have to be supplied by the

Antialigned Case

$$^{14}\mathrm{N}\Big\uparrow + \underline{\nu}_e\downarrow \rightarrow {}^{14}\mathrm{C}\cdot + e^+\uparrow$$

Aligned Case

$$^{14}\mathrm{N}\Big\uparrow + \underline{\nu}_e\uparrow \rightarrow {}^{14}\mathrm{C}\cdot + e^+\uparrow + ?\Big\uparrow$$

FIGURE 7.4 On the left is shown the spin configuration for the antialigned case, where the two spin vectors on the left hand side of the equation add up to equal the spin vectors on the right hand side. On the right, for the aligned case, the two spin vectors on the left hand side of the equation cannot add up to the spin vectors on the right hand side without additional angular momentum, indicated by the question mark which must be provided by the electron antineutrino or the positron. The dot by ^{14}C indicates that it has zero intrinsic angular momentum

electron antineutrino or the positron; it will have to come from one of their quantum mechanical wave functions. Again, these reactions are mediated by the weak interaction, and that interaction allows the antineutrino or positron to provide the one unit of angular momentum that is required here. But we know from the general properties of these reactions that they are much less probable, by roughly a factor of 10, if even one unit of angular momentum is provided by the antineutrino or positron. Thus the reaction will be much more likely for the antialigned case.

Obviously the extreme case I've described here, that is, the situation at the two throats of the neutron star, won't apply everywhere. In other locations the ^{14}N spin vectors and the neutrino spin vectors aren't just aligned or antialigned, but are at some more complicated angle. So if one is going to try to calculate the total chirality that will result from the electron antineutrino-^{14}N interactions, it will get pretty complicated. There is one more factoid that you need to know and that is, if you do all the angular momentum additions carefully, you will discover that what happens at one side of the neutron star will be exactly negated by what happens at the other. The molecular destruction at the opposite sides will leave a zero net global chirality selection unless something else happens.

But something else does happen. The strong magnetic field from the neutron star can affect the probabilities for interactions of the neutrinos and antineutrinos as they are exiting the core of the star. Indeed, several groups of researchers [9–12] have shown that this will produce a large difference in the number of electron antineutrinos at the two throats of the neutron star. Now the effects at the opposite throats of the neutron star will not cancel, and a global chirality will be produced. *Furthermore, that selectivity will be the same for every neutron star in the Galaxy, indeed, in the Universe.* We'll return to this effect a bit later.

I need to issue a few words of clarification here. Earlier we dismissed the interactions in which the electron neutrinos would convert ^{14}N to ^{14}O because the reaction threshold was about 4 MeV higher for that reaction than for the one in which ^{14}N is converted to ^{14}C by electron antineutrinos. But the former reaction will occur, and will produce exactly the opposite effect that the electron neutrinos do. It just won't be as frequent as the reaction in which the antineutrinos convert ^{14}N to ^{14}C, so the net result will be the chiral selection by the electron antineutrinos when one averages over all space. And since the conversion to ^{14}O by the neutrinos occurs in the same region of space as the conversion to ^{14}C by the antineutrinos, the latter conversion will dominate. So our dismissal of the neutrino conversion to ^{14}O was justified.

By the way, there's one other aspect of supernovae that I need to mention. Indeed, I've already noted that a supernova can outshine all the other stars in its Galaxy for a while. This is because they produce a lot of ^{56}Ni, which is radioactive, so decays to ^{56}Co, and then to ^{56}Fe, and these decays produce particles that heat their environment and cause it to shine—in photons. All this takes place with a time constant of about 80 days (the half-life of ^{56}Co is 77 days), so the light emitted from these supernovae usually falls off with that time scale (see Figure 7.5, which also shows some exceptions to the 80 day fall off). However, note that all core collapse supernovae don't emit the same amount of light (unlike the Type Ia supernovae, which we discussed in Chap. 2 as serving as the standard candles for cosmological distance determination).

All the stars shown in Figure 7.5 are shown as absolute magnitudes, that is, the amount of light one would see if one were at the same distance from each of them. Astronomers have to have a system that allows them to discuss and graph sky objects that have "luminosities" that vary by many orders of magnitude. Thus they have adopted a system of "magnitudes." The y-axis in Figure 7.5 is sort of logarithmic; one magnitude amounts to a difference in luminosity of a factor 2.512, two magnitudes to a difference of

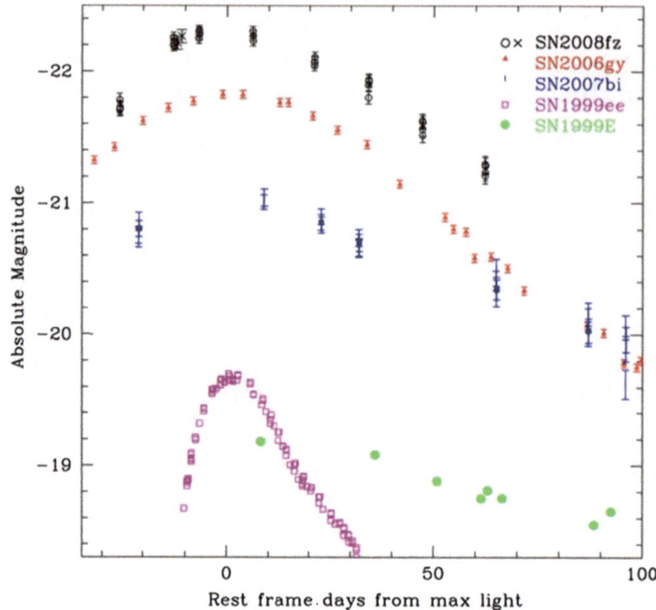

FIGURE 7.5 Comparison of the absolute magnitudes in one wavelength range of SN 2008fz with "extremely luminous supernovae" (ELSN) and normal-luminosity supernovae. Open circles: Catalina Sky Survey (CSS) [13] measurements for SN 2008fz. Crosses: measurements of SN 2008fz made with the Palomar 1.5 m telescope. Triangles are measurements for ELSN using the Katzman Automatic Imaging Telescope at the Lick Observatory for SN 2006gh [14], transformed to this frequency range. *Filled boxes* CSS measurements for ELSN, SN 2007bi [15]. *Open boxes* measurements for the normal-luminosity type Ia SN 1999ee from Stritzinger et al. [16]. *Large dots* measurements for the normal luminosity type IIn SN 1999E from Rigon et al. [17]. Figure from Drake et al. [18]. (Reproduced by Permission of the AAS. Courtesy of G. Djorgovski)

$(2.512)^2$, etc. SN 2008fz was advertised in the papers that were published about it as the brightest supernova ever observed, so it is at one extreme of the luminosity range of core collapse supernovae. What are the supernovae at the other extreme? Recall that back in Chap. 3, I mentioned silent supernovae; so some supernovae are not seen at all, at least in visible light. The interactions that the neutrinos have with the ^{14}N nuclei are pretty feeble, and in any case are much smaller than the interactions that photons would have. So it is reasonable to ask if the photons wouldn't destroy all the molecules on which the neutrinos had so carefully performed chiral selection. For many supernovae that would probably be the case.

Those silent supernovae collapse to black holes before very many photons can be emitted, but they emit all the neutrinos that the supernovae do that produce neutron stars. Heger et al. [19] studied the various modes of core collapse supernovae that can occur. They found that core-collapse supernovae occurring in stars having masses less than 25 times the mass of the Sun would ultimately produce a neutron star remnant. Those from stars ranging from 25 to 40 times the mass of the Sun will first form a neutron star, but following that, infalling matter would cause the neutron star core to exceed its maximum mass (which is supported by neutron degeneracy pressure, as noted above). It would then collapse quickly to a black hole; these are referred to as fall back black holes. A star more massive than 40 times the Sun's mass would be expected to collapse directly to a black hole.

In both the fall back black hole scenario and the direct collapse to a black hole, the time scale is short enough that much of the matter close to the black hole would be swallowed by the black hole. Of course, that matter would also trap much of the electromagnetic radiation produced in the initial explosion, so the supernova would emit its neutrinos, but much of the electromagnetic radiation would never get out. Since the ^{56}Ni would be produced close to the black hole, it would likely go into the black hole also [20]. Thus most of the black hole forming stars would emit very little electromagnetic radiation—photons—to destroy the molecules on which the neutrinos had performed their chiral selection. *These would be the supernovae that would produce the chirality selection of the amino acids that would ultimately seed the Galaxy with a preferred chirality.* But regardless of the type of supernova, the photons that are emitted either when they become supernovae, or that are emitted by their progenitor stars, are something to which we need to pay attention. We will return to the progenitor star in Sect. 7.5.

7.4 Relating the ^{14}N Spin to the Molecular Chirality

In Chap. 5, we discussed the effect of the strong magnetic field, together with the Buckingham effect, on the molecular magnetic substates. We also concluded that the Buckingham effect would drive the magnetic substates toward their maximum positive value for one chirality and toward their maximum negative value for the other. If this were all that happened, there would be no chiral selectivity; neutrino processing is essential to produce the chiral selection. The situation is illustrated in Figure 7.2. There it is seen that the electron antineutrino spin is parallel to the ^{14}N, hence maximum molecular, spin direction (when it is its molecular maximum projection) at the left hand throat of the neutron star, while the electron antineutrino spin is parallel to the ^{14}N, hence maximum molecular, spin direction (when it is its molecular maximum negative value) at the right hand throat of the neutron star. This would inhibit neutrino-induced transitions on molecules of one chirality at the left hand throat, and would inhibit transitions on molecules of the other chirality at the right hand throat.

As discussed in Chap. 5, the effect of the neutron star's magnetic field on the molecules is to redistribute the angular momentum substate populations. I have put that figure in again to remind you what it looks like, and to discuss it in more detail than we did in Chap. 5. So as the neutrinos and antineutrinos come pouring out of the neutron star, there will be selective destruction of the molecules in which the ^{14}N spin is anti-aligned with the spin of the electron antineutrinos. This will preferentially destroy the least populated states of the left handed molecules (see Figure 7.6) and the most populated states of the right handed molecules at

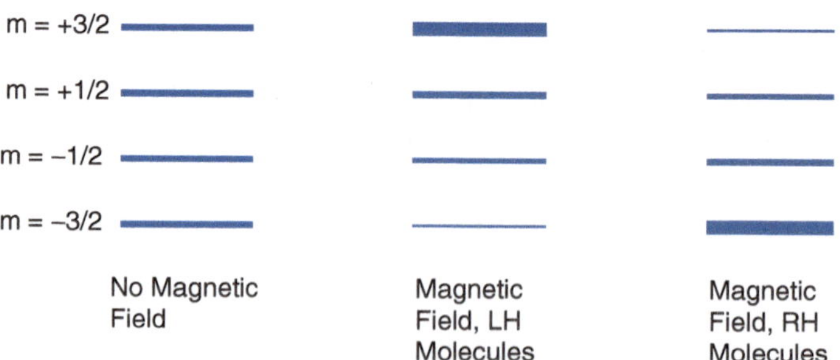

FIGURE 7.6 Populations of the different magnetic substates for an atom having angular momentum J=3/2 in a zero magnetic field (*left*), and for left-handed molecules (*center*) and right-handed molecules (*right*) in the presence of an external magnetic field. The thickness of the line indicates the magnitude of the population of each substate. (From Boyd et al. [2]. Courtesy of the International Journal of Molecular Sciences)

the left-hand throat of the neutron star, thereby producing a chiral selection of the left handed molecules at that throat.

And, as noted above, exactly the opposite effect occurs at the right-hand throat of the neutron star. But as previously mentioned, it has been shown by several theoretical groups [9–12] that the incredibly strong magnetic field of the neutron star will affect the probabilities for the neutrinos to interact with the matter that exists as the neutrinos are departing the neutron star. The neutrinos don't interact very much, but the matter in a neutron star is extraordinarily dense, so the neutrinos will interact a few times before they exit the star. However, the result of these interactions in the magnetic field of the neutron star is that there is an asymmetry in the number of neutrinos emitted parallel and antiparallel, that is, in the left-hand direction and in the right-hand direction in Figure 7.2. The difference may be as large as 30%. And it is this effect that destroys the delicate balance that exists between the chirality selection effects of the left handed molecules at the left-hand throat of the neutron star and right handed molecules at the right-hand throat of the neutron star, and produces a net global chirality for the amino acids.

7.5 The Supernova Progenitor Star

We also need to be concerned about the kinds of stars that can produce the neutron stars and black holes that will perform the chiral selection this model requires, and how they would affect their environment before the supernova occurred. Clearly the stars must be sufficiently massive that they become core-collapse supernovae; the attendant production of the electron antineutrinos is essential. A magnetic field is also crucial for providing the induced magnetic moment in the molecules and orienting the ^{14}N spin.

As noted above, the stars that will ultimately become supernovae have a downside. Stars that have masses of roughly 8–25 times the mass of the Sun will, in their helium burning phase of stellar evolution (see Chap. 3), become "red giants." After the hydrogen in the core of the star has been consumed and the star begins to burn its helium, the energy that is produced is so enormous that the outer portions of the star are forced out by the pressure of the radiation being produced in the core. The star becomes much larger than it was during its hydrogen burning phase, and because it is larger, the energy output per square centimeter per second of the surface is reduced, even though the total energy output per second has increased. This will make the surface appear to be redder than it was for the star in its hydrogen burning phase, thus suggesting the name of the resulting star.

The radius of a typical red giant is so much larger than it was when the star was in its previous (hydrogen burning) phase that it will typically extend to a distance that is larger than the orbit of the Earth around the Sun. (And, yes, when our Sun goes into its red giant phase it will envelope the Earth. But don't rush out to buy life insurance; that's several billion years down the road.) It will also be somewhat beyond the maximum radius at which the neutron star's magnetic field will provide the alignment necessary for the neutrinos to process molecules. The molecules will be *inside* the red giant! That, of course, would be likely to destroy the molecules, unless they are shielded in some way from the intense heat from the progenitor star.

What about the stars with masses greater than 25 solar masses? Heger et al. [19] found that these stars are likely to become Wolf-Rayet stars, which shed one or two of their outer layers in stellar winds that are the result of the extreme radiation pressure from the cores of these massive stars. Indeed, the products of the nucleosynthesis that occur in the outer shells can be seen in the resulting clouds: nitrogen, the dominant element resulting from hydrogen burning, and in more massive Wolf-Rayet stars, carbon, the result of helium burning (see Chap. 3). Furthermore, the stellar winds are sufficiently dense that they often obscure the central object [21]. When astronomers look at a star with a "spectrograph" (see the description in Chap. 4), they can determine the intensities of the different constituents of the "photosphere," or outer portion of the star, from their absorption lines. To produce these, the hot inner star "back lights" the outer part, which absorbs the particular wavelengths of lines that are characteristic of the elements therein. But Wolf-Rayet stars are different; they produce emission lines that characterize those elements in their winds—the nitrogen or the carbon that the astronomers observe—and completely obscure the emissions from the star itself.

Another interesting feature is that dust grains have been observed in these clouds (see, for example, refs. [21, 22]). They are apparently made when the wind from a Wolf-Rayet star collides with the wind from a massive companion star. However, this scenario would not be able to provide the molecule-containing grains on which our model depends for the following reasons: First, although binary systems are common, very massive stars are not, and binary systems with two massive stars are extremely rare. The second problem is that the grains are thought to occur only beyond several tens of AU (AU = astronomical unit = the distance from the Earth to the Sun = 1.5×10^{13} [15 trillion] centimeters) [21] from the Wolf-Rayet star, and the magnetic field from the nascent neutron star, when it forms, cannot align the molecules beyond a distance somewhat less than one AU [1]. The third problem is that, despite the fact that the Wolf-Rayet stars are quite small (less than the size of the Sun) they are extremely hot (their surfaces are, after all, much closer to the sources of the stellar energy than is the situation in a star that has not shed its outer layers), so even though the material in the winds from the star absorbs a lot of the energy from the star, the temperature is still too high, even as far away as tens of AU–for the grains to form.

Thus, the grains that form in the Wolf-Rayet cloud are still in an extremely high temperature region; these would certainly be too hot to sustain formation of molecules. One could imagine the winds from the surface of a Wolf-Rayet star coming off in blobs [23], which might help in shielding the inner dust grains that formed from the intense heat of the star, and there is evidence that this does happen [24], but even this seems to require extreme assumptions to save the dust grains from destruction.

I mentioned previously that there was a maximum distance at which the neutrino processing can occur for this single supernova. In the SNAAP model, this will be determined by the point at which the magnetic field from the nascent neutron star or black hole is no longer aligned, along the collapsed star's magnetic axis, with the direction of the neutrinos. This will happen because there is a Galactic background magnetic field; it is about 10^{-6} gauss—one millionth of a magnetic field unit [25] (the Earth's magnetic field is roughly one gauss). So when the magnetic field from the neutron star or black hole becomes comparable to this background magnetic field, that direction can no longer be maintained, and the selective processing resulting from the alignment due to the magnetic field and the neutrino chirality can no longer take place.

Therefore, it is not looking good for having molecules so they can be processed by this combination of neutrinos and magnetic fields. By the time a grain gets far enough from the star that it could sustain creation of molecules that wouldn't be destroyed as soon as they formed, there isn't a strong enough magnetic field to provide the orientation for the chiral processing. However, this would not be the case for dust grains or meteoroids that just happened to be passing through the cloud. There are such objects flitting about the Galaxy all the time, and however many of them are within one AU of the star when it becomes a supernova will be processed. Indeed this seems to be the most likely scenario for chirality selection. These, of course, would be meteoroids that had been produced somewhere else, so many of them would have had time to produce and replicate molecules. These would also be racemic unless they had already been near a supernova when it exploded, in which case the current supernova would just increase the level of enantiomerism.

So if a meteoroid happened to be passing by a bit closer than one AU of the star when the supernova went off, its molecules would be processed. And there are bound to be quite a few meteoroids passing by any time a supernova goes off. Remember, we are talking about a volume that has a radius of the distance from the Earth to the Sun, and that is huge. During meteorite shower peaks, we Earthlings see a wonderful streak in the sky every minute or so; that's in a volume that is ten trillion times smaller than the volume surrounding a Wolf-Rayet star in which the molecules would be processed! And the volume that we Earthlings sample has also been swept pretty clean by Jupiter, whereas one would not expect the volume around the Wolf-Rayet star to have been subjected to the Jovian janitor.

But there's one more thing we need to worry about, that being the high radiation field that the meteoroids and grains will encounter as they pass through enough of the Wolf-Rayet star to get them within range of the magnetic field and the neutrinos from the supernova. This radiation is sufficiently intense that it would probably completely vaporize small grains, and that would surely evaporate any molecules from the surfaces of larger grains. But this would not be the case for meteoroids that were agglomerations of smaller grains, but had some molecules on their surfaces before they got agglomerated into the larger meteoroid. In this case, some of the surface material can be ablated away while still leaving inner material—and molecules. Indeed, the assumption of larger objects passing by may also relax the conditions on the progenitor star—the red giant—for supernovae from less massive stars, that is, those that ultimately produce a neutron star. However, it would probably not allow even very large objects to survive the incredible photon blast from a supernova that did not swallow its photons in its subsequent collapse to a black hole, so the Wolf-Rayet star is probably the only one that will work for the SNAAP model.

Can we estimate what the efficiency of chirality selection would be? We can assume that the closest distance that a grain or meteoroid could come to a Wolf-Rayet star, presumably containing the molecules of interest (created elsewhere), without having its molecules destroyed, is about 10^{12} cm. We can guess what the "cross section" (the probability for one neutrino on one nucleus to undergo the reaction of interest) would be from theoretical studies done on other nuclei [26] for antineutrinos undergoing a charged-current weak interaction on ^{14}N; that interaction probability might be around 10^{-40} square centimeters. This incredibly small number is as small as it is for two reasons. First, it requires that the neutrino hit the nucleus, which looks to the incident neutrino like a circle with a radius of about 3×10^{-13} cm (For ^{14}N, that's three ten thousandths of a billionth of a centimeter.), so has an area of about 3×10^{-25} square centimeters. If the incident particle were a proton instead of a neutrino, it would interact via the strong interaction, and would have a cross section of about that same size. This is to say that if the proton hit the nucleus (and had enough energy) it would perform some kind of a nuclear reaction. That's not the case with the neutrino, though, which only interacts via the weak interaction; that accounts for the additional factor of 3×10^{-14} in the cross section.

Getting back to our estimate of the efficiency, we can assume that there are about 2×10^{56} electron antineutrinos emitted when the supernova explodes, which then gives the probability of any molecule being destroyed by this interaction to be about 2×10^{-9}, two in a billion! This is incredibly small, although it would be larger if we were considering a larger object, for example, a significant meteoroid that happened to wander by as the supernova went off. For such an object, the distance from the star could be smaller, as the inner part of the meteoroid could be shielded by the outer part. The neutrinos would, of course, process the entire object. If the object were a factor of 10 closer to the star, its probability of interaction would increase by a factor of 100. In any event, from there, amplification would take over to increase the enantiomeric excess.

Indeed, it is not at all unreasonable to assume that a meteoroid could be within 10^{11} cm of the star; that's a bit less than 1/100 of an AU. The meteoroid would have to be initially large enough and be moving sufficiently swiftly so that some of it could be ablated away while protecting the part that remained. It would also have to not be in the high-temperature region of the central star for long enough to have all of its remaining molecules destroyed. If that is the case, now the probability of the molecules in that dust grain or meteoroid being processed are improved: 2×10^{-7}; one in five million! I've included a Figure 7.7, that shows the distance scales that are relevant to this discussion.

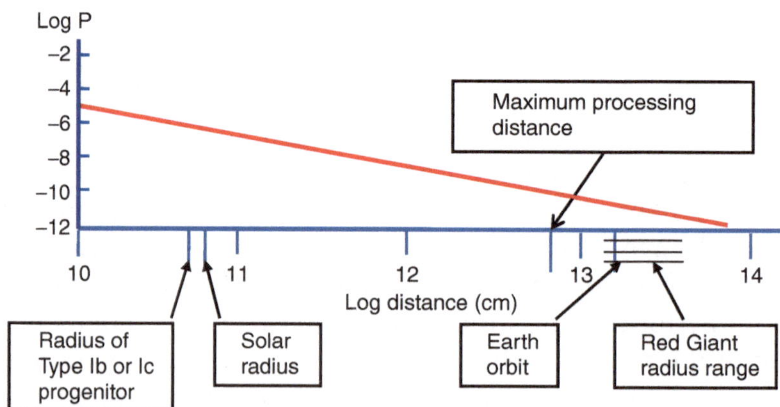

FIGURE 7.7 Distances relevant to processing grains or meteoroids in the magnetic field and electron antineutrino flux of a supernova. The vertical axis gives an estimate of the processing probability as a function of distance.

There are some other interesting factors that were not included in this calculation. These were discovered by Lunardini [27], who found that the fraction of electron antineutrinos for the very massive supernovae was increased from that for the less massive supernovae. She also found that their mean energy was increased. Both of these factors would serve to increase the processing efficiency. I'll return to this in the next chapter.

So the supernovae from the massive stars can facilitate the processing that we need to produce the enantiomerism in the amino acids, but the level of enantiomerism is quite small. It is probably larger than the net enantiomerism produced in other models, but it will certainly require some amplification, either in the interstellar medium, or once the molecules arrive on a planet, or most likely, as noted in Chap. 6, in both situations.

References

1. R.N. Boyd, T. Kajino, and T. Onaka, Supernovae and the Chirality of the Amino Acids, Astrobiology 10, 561 (2010)
2. R.N. Boyd, T. Kajino, and T. Onaka, Stardust, Supernovae, and the Chirality of the Amino Acids, Int. J. Mod. Sci. 12, 3432 (2011)
3. A. Esteben-Pretel, S. Pastor, R. Tomas, G.G. Raffelt, and G. Sigl, Multi-angle Effects in Collective Supernova Neutrino Oscillations, Phys. Rev. D 767, 125018 (2007)
4. M.Th. Keil, G.G. Raffelt, and H.T. Janka, Monte Carlo Study of Supernova Neutrino Spectra Formation, Astrophys. J. 590, 971 (2003)
5. H. Duan, G.M. Fuller, J. Carlson, and Y.-Z. Qian, Simulation of Coherent Nonlinear Neutrino Flavor Transformation in the Supernova Environment: Correlated Neutrino Trajectories, Phys. Rev. D 74, 105014 (2006)
6. H. Duan, G.M. Fuller, J. Carlson, and Y.-Z. Qian, Analysis of Collective Neutrino Flavor Transformation in Supernovae, Phys. Rev. D 75, 125005 (2007)
7. H. Duan, G.M. Fuller, and Y.-Z. Qian, Simple picture for Neutrino Flavor Transformation in Supernovae, Phys. Rev. D 76, 085013 (2007a)
8. J. Gava, J. Kneller, C. Volpe, and G.C. McLaughlin, Dynamical Collective Calculation of Supernova Neutrino Signals, Phys. Rev. Letters 103, 071101 (2009)

9. C.J. Horowitz and G. Li, Cumulative parity violation in supernovae. Phys. Rev. Letters 80, 3694 (1998)
10. D. Lai and Y.-Z. Qian, Neutrino transport in strongly magnetized proto-neutron stars and the origin of pulsar kicks: the effect of asymmetric magnetic field topology. Astrophys. J. 505, 844 (1998)
11. P. Arras, and D. Lai, Neutrino-nucleon interactions in magnetized Neutron-star matter: the effects of parity violation. Phys. Rev. D 60, 043001-1 (1999)
12. T. Maruyama, T. Kajino, N. Yasutake, M.-K. Cheoun, and C.-Y. Ryu, Asymmetric Neutrino Emission from Magnetized Proton-Neutron Star Matter Including Hyperons in Relativistic Mean Field Theory. Phys. Rev. D 83, 081303-1 (2011)
13. S.M. Larson, for a description of the Catalina Sky Survey project see http://stardust.jpl.nasa.gov/news/news17.html, updated November 26, 2003
14. N. Smith, W. Li, R.J. Foley, J.C. Wheeler, D. Pooley, R. Chornock, A.V. Filippenko, J.M. Silverman, R. Quimby, J.S. Bloom, and C. Hansen, SN 2006gy: Discovery of the Most Luminous Supernova Ever Recorded, Powered by the Death of an Extremely Massive Star like η Carinae, Astrophys. J. 666, 1116 (2007)
15. Gal-Yam, A., P. Mazzalli, E.O. Ofek, P.E. Nugent, S.R. Kulkarni, M.M. Kasliwalk, R.M. Quimby, A.V. Filippenko, s.B. Cenko, R. Chornock, R. Waldman, D. Kasen, M. Sullivan, E.C. Beshore, A.J. Drake, R.C. Thomas, J.S. Bloom, D. Poznanski, A.A. miller, R.J. Foley, J.M. Wilverman, I. Arcavi, R.S. Ellis, and J. Deng, Supernova 2007bi as a Pair-Instability Explosion, Nature 462, 624 (2009)
16. M. Stritzinger, M. Hamuy, N.B. Suntzeff, R.C. Smith, M.M. Phillips, J. Maza, L.G. Strolger, R. Antezana, L. Gonzalez, M. Wischnjewsky, P. Candia, J. Espinoza, D. Gonzalez, C. Stubbs, A.C. Becker, E.P. Rubenstein, and G. Galaz, Optical photometry of the Type Ia SN 1999ee and the type Ib/c SN 1999 ex in IC 5179, Astron. J. 124, 2100 (2002)
17. L. Rigon, M. Turatto, S. Benetti, A. Pastorello, E. Cappellaro, I. Aretxaga, O. Vega, V. Chavushyan, F. Patat, I.J. Danziger, M. Salvo, SN 1999E: Another Piece in the Supernova-Gamma-Ray Burst Connection, Mon. Not. Royal Astron. Soc. 340, 191 (2003)
18. A.J. Drake, S.G. Djorgovski, J.L. Prieto, A. Mahabal, D. Balam, R. Williams, M.J. Graham, M. Catalan, E. Beshore, and S. Larson, Discovery of the Extremely Energetic Supernova 2008fz, Astrophys. J. 718, L127 (2010)
19. A. Heger, C.L. Fryer, S.E. Woosley, N. Langer, and D.H. Hartmann, How Massive Stars End Their Life, Astrophys. J. 591, 288 (2003)

20. C. Fryer, Neutrinos from Fallback onto Newly Formed Neutron Stars, Astrophys. J. 699, 409 (2009)
21. P.A. Crowther, Physical Properties of Wolf-Rayet Stars, Ann Rev. Astron. Astrophys. 45, 177 (2007)
22. P.M. Williams, K.A. Van der Hucht, and P.S. The, Infrared Photometry of Late-Type Wolf-Rayet Stars, Astron. Astrophys. 182, 91 (1987)
23. S. Lepine, A.F.J. Moffat, N. St-Louis, S.V. Marchenko, and M.J. Dalton, P.A. Crowther, L.J. Smith, A.J. Willis, I. Igor, and G.H. Tovmassian, Wind Inhomogeneities in Wolf-Rayet Stars, IV. Using Clumps to Probe the Wind structure in the WC8 Star HD 192103, Astron. J. 120, 3201 (2000)
24. A.F.J. Moffatt, L. Drissen, R. Lamontagne, and C. Robert, Spectroscopic Evidence for Rapid Blob Ejection in Wolf-Rayet Stars, Astrophys. J. 334, 1038 (1988)
25. K.M. Ferriere, The Interstellar Environment of our Galaxy, Rev. Mod. Phys. 73, 1031 (2001)
26. G.M. Fuller, W.C. Haxton, and G.C. McLaughlin, Prospects for Detecting Neutrino Flavor Oscillations, Phys. Rev. D 59, 085005 (1999)
27. C. Lunardini, Diffuse Neutrino Flux from Failed Supernovae, Phys. Rev. Letters 102, 231101–1 (2009)

8. Spreading Chirality Throughout the Galaxy and Throughout the Earth

Abstract To what extent would the molecules that have been processed by SNAAP become widely distributed throughout the Galaxy? As we discussed in earlier chapters, the molecules are thought to be created and contained, probably in dust grains, possibly in larger objects such as meteoroids or even comets (which could be agglomerations of grains) in the molecular clouds that pervade the Galaxy. The objects must then come relatively close to a star when it becomes a supernova so that they can be processed by the supernova neutrinos. Would this model predict the creation of a homochiral environment on Earth? Would it create the same chirality in every planetary environment? This chapter revisits amplification in these contexts, and presents an experimental test for the SNAAP model. It also revisits the cometary missions, and predicts the results from them for the SNAAP model and the circularly polarized light model.

8.1 Might Supernovae Process All the Amino Acids in the Galaxy?

The supernovae by themselves would not be able to perform all of these tasks (it's not even close!), as I'll try to show with a simple calculation, although it involves multiplying a bunch of really large numbers. If you'd just as soon take my word for it, just skip the rest of this paragraph and fast-forward to the following one. The frequency of core-collapse supernovae is roughly one every 30 years in the Galaxy [1], or 3×10^5 (three hundred thousand)

R.N. Boyd, *Stardust, Supernovae and the Molecules of Life: Might We All Be Aliens?*, Astronomers' Universe, DOI 10.1007/978-1-4614-1332-5_8,

during the lifetime of a molecular cloud [2, 3], roughly ten million years, assuming all those supernovae are in the same molecular cloud (obviously an extreme assumption). The fraction of these supernovae that go to black holes is not well known, but may be of order 20% [1, 4]. The magnetic dipole field falls off as $1/r^3$, and neutron star magnetic fields are thought to be around 10^{14} (one hundred trillion) gauss or more at their surface [5] of radius ~10 km. The cloud's molecules would be oriented by the magnetic field of the neutron star if they were within 5×10^{12} (five trillion) centimeters of the center of the supernova, or perhaps a bit more, so the volume processed by the 3×10^5 supernovae, assuming their processing volumes do not overlap (another extreme assumption, and in the same direction as the other one), and that they all occurred in the same molecular cloud, would be 1×10^{43} cm^3 (I have to use scientific notation here; anything else would be prohibitively cumbersome.). Roughly 2×10^9 (two billion) Solar masses of material are in the Galactic molecular clouds [6], at a density ranging from 10^2 to 10^6 (one hundred to one million) molecules or atoms per cubic centimeter—assuming 10^4 (ten thousand) per cubic centimeter—giving a total cloud volume of 2.4×10^{62} cm^3.

Thus the volume of a cloud is far greater than the volume that could be processed by the supernovae that would occur during the cloud's lifetime, even with the extreme assumptions that we made [7, 8]. So, although the supernovae do produce the electron antineutrinos and the magnetic fields for some of the molecular processing, they clearly need to rely on amplification mechanisms, possibly both in the interstellar medium and again once the chiral molecules begin their planetary existence, to finish the job of producing homochirality.

We just concluded that the collapse of massive stars to black holes cannot possibly provide for the amount of Galactic processing needed to spread enantiomerism throughout the Galaxy, but we should note that we may have underrated the capabilities of the massive stars (although this will not affect this conclusion). As was observed in the previous chapter, Lunardini [4] has studied the neutrinos that would be emitted from a star that was sufficiently massive that it would collapse to a black hole. She found that the fraction of the neutrinos that emerged from such massive stars that were electron antineutrinos was higher than has been estimated for less massive stars, and furthermore that their energy would be higher than it would be for less massive stars. This would have a twofold effect, and they both enhance the efficiency that the electron antineutrinos would have in selecting a chirality. The additional neutrinos would permit processing of molecule containing grains with a higher probability than would be permitted by the lower neutrino fluxes. Furthermore, the reaction probability, which varies as the square of the energy over the threshold energy, would be increased by the increased energy. Thus our estimate of the processing efficiency of the supernovae may well be an underestimate. Indeed, the processing efficiency at a distance of 10^{11} cm from the Wolf-Rayet star may increase to as much as 10^{-6}; one in a million. Although this is about the maximum value that the processing efficiency a single supernova could be expected to have anywhere (it would be smaller farther out from the supernova), this is the most relevant value; it is the most likely point at which the amplification would begin!

8.2 Spreading Molecules and Enantiomerism Throughout the Galaxy

Although we found that the supernovae cannot seed the entire Galaxy, or even a single molecular cloud, with more than a tiny level of enantiomerism, it may be quite possible for them to first produce that tiny amount of enantiomerism, then enlist a combination of chemical evolution and Galactic mixing to amplify the enantiomerism and mix it over a large fraction of the Galactic volume. Indeed we [7, 8] believe that these mechanisms are unavoidable, based on the work of a lot of other experimental and theoretical scientists. As soon as a supernova explodes, the material it has produced with a favored chirality will begin to mix with the smaller molecules of the molecular cloud to produce more molecules, having the same favored chirality, either by the autocatalysis mechanism described by Gol'danskii and Kuz'min [9] (see Chap. 6) or the replication process described by Garrod [10] or Hasegawa et al. [11], or perhaps some combination of those mechanisms.

8.2.1 Cosmic Amplification

For example, as was discussed in Chap. 6, Garrod et al. [10] and Laas et al. [12] studied the possibility that replication of complex molecules would occur in the interstellar medium in the warmed ice outer shells of grains, possibly even on surfaces of grains that were part of an agglomeration of grains if the object were sufficiently porous (as carbonaceous chondrites appear to be). Chemical replication in the study of Garrod et al. [10] could be catalyzed by radicals, for example, H, OH, CO, CH_3, NH, and NH_2, created by the interactions of high energy cosmic rays with preexisting molecules. Would this replication preserve and amplify any existing enantiomeric excess? The models of Gol'danskii and Kuz'min [9], and of Kondepudi and Nelson [13] suggest that it could, and they give the criteria for this to occur, although without any details about the structure of specific molecules. However, the models do seem to describe a viable mechanism for amplification.

In any event, it does seem plausible that either autocatalysis, or chemical replication via radicals, or both, could produce the chemical replication required to amplify the chirality established by the magnetic fields and neutrinos from supernovae. This is fortunate; a large amplification factor within the cosmic environment appears to be an essential feature of the SNAAP model, indeed of any of the models proposed to explain the chirality of the amino acids, since all models produce maximum values of their enantiomeric excesses that are very tiny, at least on a global scale.

However, any of the models for amplification need to face the question: is one part in a million (or much less in some of the other models) enough to trigger a subsequent large enantiomerism, for example, the enantiomeric excesses of order 10% (or perhaps even higher) found in the Murchison meteorite? The laboratory experiments that have looked at amplification have certainly achieved large enantiomeric excesses, but they started with enantiomeric excesses around 1%. What would happen if they started with one part in a million? Would they be able to even measure the "progress" in their experiment over the lifetime of the scientists involved? This question should provide a rich test bed for clever experimental chemists, and a lot of Ph.D. thesis. The problem, of course, is that these experiments might take many years, and (well justified) graduate student impatience may not be compatible with that time frame!

8.2.2 Galactic Mixing

If the enantiomeric excess of the processed material can be increased via chemical replication, it would also be expected to mix with a much larger fraction of the material in the Galaxy, ultimately establishing at least a preference for left-handed amino acids throughout a large fraction of the Galaxy, or at least within the volumes occupied by the molecular clouds. Details of the processes by which this might occur have been discussed by Pittard [14] in exquisite detail. He discusses the mechanisms by which mass, momentum, and energy are exchanged between diffuse plasmas that include "many types of astronomical sources, including planetary nebulae, wind-blown bubbles, supernova remnants, starburst superwinds, and the intracluster medium." Let me give a bit more definition to these terms:

- Planetary nebulae: These are the glowing shells of ionized gas and plasma usually ejected during the asymptotic giant branch (helium burning) phase of certain types of stars. For more information, go to http://en.wikipedia.org/wiki/Planetary_nebula.
- Wind-blown bubbles: These result from the winds from massive (greater than eight Solar mass) stars. See, for example, Dwarkadas [15] for more information.
- Supernova remnants: These are comprised of the gas that contains the newly synthesized elements that supernovae expel into the interstellar medium. These alone would expand the volume processed during the time of the supernova by a huge factor. More information can be found at http://en.wikipedia.org/wiki/Supernova_remnant.
- Starburst superwinds: These are winds that are generated from regions with an unusually high density of supernovae—the starburst regions. However, the superwinds are several times greater in mass than could be generated by the supernovae alone, suggesting that there is additional evaporation of material from the cloud that houses the starburst region. For more information, see, for example, Hartquist et al. [16].

- Intracluster medium: This is the superheated gas near the center of a galactic cluster. It contains mostly hydrogen and helium, and strongly emits x-rays. For more information, see http://en.wikipedia.org/wiki/Intracluster_medium.

Pittard also notes "The injection and mixing of mass from condensations into a surrounding supersonic medium induces shocks, increasing the pressure of the flowing medium. This can [result] in star formation, and is likely to play a role in sequential star formation..." Suffice it to say that, with the dynamical processes that exist in the Galaxy, redistribution throughout the molecular clouds, and ultimately throughout much of the Galaxy, would unavoidably occur. This mixing would be expected to occur on a much slower timescale than the chemical processing timescale, as discussed below, but would inevitably establish a preferred chirality throughout much of the Galaxy, *provided the same chirality was consistently selected by the mechanism by which the preferred chirality was established.*

It is important to consider the timescales of the chemical evolution and the galactic mixing for the SNAAP model to produce some enantiomerism throughout at least some of the Galaxy to ascertain that these occur on timescales less than the age of the universe. Although abundance inhomogeneities surely do exist within the Galaxy, it is generally assumed that the Galactic mixing timescale is much smaller than the age of the universe [17, 18]. As one signature of Galactic mixing time, our Galaxy rotates roughly once every 3×10^8 years (300 million years; see, for example, Ref. [19]), much less than the $\sim12\times10^9$ (12 billion) years the Galaxy has lived. The evolutionary timescales of organic molecules are undoubtedly much shorter than the Galactic mixing timescales. Although this might depend on many variables, the fact that such molecules are observed by astronomers to have been created in the molecular clouds, and that these clouds are born, live, and die in of order ten million years (see, for example [2, 3]), confirms the shortness of the chemical evolutionary timescale.

8.2.3 Planetary Amplification

As discussed in Chap. 6, the ability of a collection of molecules exhibiting a small enantiomeric excess to amplify that excess dramatically under conditions that could well be replicated in planetary environments (if they had surface water!) has been demonstrated (see, e.g. [20–22]). Furthermore, as observed by Breslow and Levine [23], as soon as biologically active systems begin to be involved, the drive toward homochirality can proceed very rapidly, although, as noted by Gol'danskii and Kuz'min [9], at least local homochirality may be essential for biological synthesis to proceed at all. None the less it does appear that a barrage of meteorites that had slightly enantiomeric amino acids could trigger a second phase of amplification that might well be expected to drive the somewhat enantiomeric molecules they contained to homochirality very quickly once they were delivered to a planet by a cosmic stork.

8.3 Testing the Models of Amino Acid Chirality

While it might be thought to be unnecessary to speculate about the enantiomerism that this model would spread throughout the Galaxy, based on the tiny number of meteoritic samples (which also contained collections of racemic molecules), and the number of Earthly living beings (although there are billions of us, we may only count as one datum, since the only relevant datum may be the common source of our amino acid homochirality), the SNAAP model for propagating the supernova-selected enantiomerism throughout the Galaxy allows for only one chirality to exist, at least on a global scale. Thus this constitutes an important prediction of the model; finding a statistically significant abundance preference for right-handed amino acids would be very difficult for it to explain.

As noted in Chap. 4, one new test may occur before long; the Japanese space mission Hayabusa [24–26] returned to Earth in 2010, and may have some samples from asteroid Itokawa; the amino acids included in those samples must have predominantly left-handed chiralities. When these samples have been analyzed for their amino acids, assuming it can be established that they did indeed come from asteroid Itokawa (and this is not trivial; it is very easy to contaminate the sample, especially when the amount of material is very small), this information should provide a strong test of the SNAAP model. And the ROSETTA mission [27] should provide another test following the arrival of its lander on comet Churyumov Gerasimenko in 2014, or shortly thereafter, since it takes time to analyze data and for the members of the collaboration to be sure they all agree on the interpretation of the results.

These tests are crucial to all the models discussed in this book. In the context of the SNAAP model, if some clumps of amino acids are found by the ROSETTA mission to be right-handed, it would not be fatal for the model. A possible scenario for the SNAAP model to produce right-handed molecules would involve some of the meteoroids being processed on the opposite side of the neutron star (or black hole) that selected left-handed molecules, ending up with opposite chiralities for its molecules. It is because, as discussed in Chap. 7, the electron antineutrino fluxes from the supernovae are not equal on the two sides of the nascent neutron star that the ones produced by converting ^{14}N to ^{14}C would still dominate over all space. As noted in Chap. 7, if mixing was not adequate to stir the left-handed molecules from one side of the neutron star with the right-handed molecules from the other side, enough right-handed molecules would be produced to generate some right-handed entities. So the SNAAP model still could survive if such molecules were found in a single comet, although it is somewhat difficult to imagine how the galactic mixing mechanisms discussed above could accommodate this.

If one could measure the ratio of right-handed amino acids to left-handed amino acids in a pristine environment, it would produce a very significant test of all the models I've discussed (although, in actuality, that might turn out to be a test of the completeness of the mixing that occurred). For the SNAAP model, the global enantiomeric excess could be estimated if the electron neutrino and electron antineutrino energy spectra could be accurately determined, although as noted above, the most relevant enantiomeric excess may be the maximum value that a single supernova can produce.

Recall that in Chap. 5 we discussed the circularly polarized light model described by de Marcellus [28] that assumed that the material of which the Solar System is made was bathed, at some point in its past, by uniformly circularly polarized light, and that this would produce the same chirality of the amino acids throughout the Solar System. If this model is correct, it would not allow for any right-handed amino acids. Of course, Galactic mixing prior to formation of the Solar System might dilute the effect, in which case some right-handed amino acids might then be found.

It might be possible to test this idea if we could locate a "hyperbolic comet," which would have been formed outside the Solar System, and would make one pass through before returning to outer space, on a hyperbolic trajectory, and test some of its material. If such an object contained amino acids with the opposite chirality from those in the Solar System, and the amino acids from Hayabusa and Rosetta were all left-handed, it would lend credence to the model of de Marcellus. The circularly polarized light that processed the material of the Solar System would have to process material somewhere else with the opposite polarization. Although that material could have been processed by circularly polarized light with the same chirality as that assumed to have processed the Solar System, thereby producing amino acids with the same chirality as those of the Solar System, it might also have been processed, far from the Solar System, with circularly polarized light of the opposite chirality, thus producing right-handed amino acids. The unfortunate aspect of this test is that hyperbolic comets are extremely rare. But then, no one ever guaranteed that science would be easy!

You have no doubt noticed that, while I have described a variety of experiments that have been performed to test the variety of models of amino acid chirality that I have described, I have not presented much, if any, data in support of the SNAAP model. The probabilities for neutrinos to interact with matter are extremely small, so such data require pretty heroic experiments to acquire any data. And, of course, neutrino "sources" are really particle accelerators, and not little ones either. Neutrino nucleus interaction probabilities, "cross sections" to physicists, have been measured for a number of nuclei, but no one has tried to infer how they might affect the chiralities of the amino acids. However, I believe experiments might be conducted to at least test some of the ideas that are basic to the SNAAP model.

Probably the best place to do an experiment, both because the neutrino beam would be quite intense there, and because the energies of the neutrinos would be roughly comparable to those that would be emitted by a supernova, would be the Spallation Neutron Source at Oak Ridge National Laboratory in Tennessee. This facility was built for the purpose of producing a neutron beam to be used in studying neutron induced processes, but it also will produce an intense neutrino beam. The primary driver for this facility is a medium energy proton beam, which will produce copious numbers of positively charged pions (see Chap. 7) through its interactions with target nuclei. These will decay to muons and a muon antineutrino, which will then decay to a positron and two more neutrinos, and the positron will annihilate with an electron to produce photons. The resulting neutrino spectrum is shown in Figure 8.1. As can be seen, the neutrino energies, tens of MeV, are at least similar to those from a supernova, shown in Figure 7.3. Also shown in Figure 8.1 is the time distribution of the different neutrino flavors that will be produced by the Spallation Neutron Source. This is especially important in distinguishing between events initiated by the different flavors of neutrinos, or between a neutrino and its own antineutrino.

So what would an experiment to test the SNAAP model look like? The most obvious experiment would be one in which a

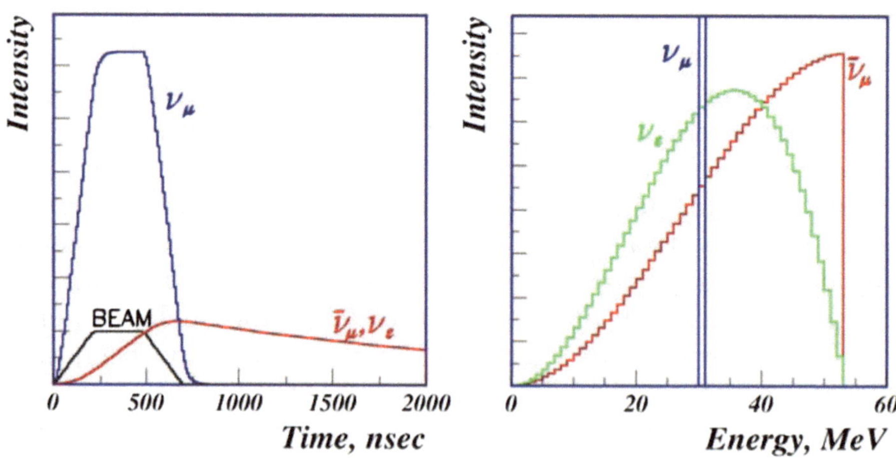

FIGURE 8.1 Arrival times of the different neutrino flavors and the spectrum of neutrino energies produced by the Spallation Neutron Source. "nsec" = nanoseconds = billionths of a second. (Courtesy of Y. Efremenko. From [29], published by IOP Publishing, LTD)

neutrino beam would bombard a vat of amino acids, which would be encased in as strong a magnetic field as could be generated with a laboratory magnet. Somehow the conversions of the nitrogen nuclei to carbon or oxygen would be observed. It would also be important to reverse the magnetic field at some point to see if the reaction rate would be affected (the neutrino spin direction would remain the same, so this would simulate the situation at the opposite end of the neutron star). Unfortunately, the experiment might be complicated by the need to simulate the motion of the meteroids in the magnetic field of the neutron star. This might necessitate either rotating the amino acid sample on a large wheel, half of which would be shielded from the magnetic field, and half would be subjected to the field of the magnet (it's not easy to shield the target from the neutrinos!). Or it might involve flowing a liquid that included the amino acids through the magnetic field region.

Of course, the SNAAP model also could operate together with some model that would allow both chiralities to exist in approximate equality, for example, the model involving circularly polarized light. Although the SNAAP model seems to be winning for the moment, with a dominant chirality, more statistics on the chirality of the amino acids in the cosmos, for example, as will be provided by the data from the comets, will tell us a lot about the validity of the different theories. And, of course, it would be important to obtain data from a comet or two containing material from outside of the Solar System, that is, from the previously mentioned hyperbolic comets. However, if a lot of the new data showed that there were more right-handed amino acids than left handed ones, the SNAAP model, and every other model I've discussed, would be in trouble. But every scientific theory has this same potential fate. One can never "prove" a scientific theory to be correct, but only to show that it has passed the most recent test. Such is the nature of science!

An experiment such as this would be important for another reason. Technically we do not know if the SNAAP model actually would favor left-handed amino acids, only that it would favor one of the two possible chiralities. Although detailed calculations might shed light on this issue, it would be much more convincing to do an experiment that would determine unequivocally which chirality was selected. The SNS experiment would do that, and for more than a single amino acid.

8.4 The Impact of Amino Acid Chirality on Questions of the Origin of Life in the Universe

As noted in Chap. 1, a period of intense meteorite bombardment on the Moon, and presumably also on Earth, occurred early in the Earth's history. Although it is not clear when this began, it does appear to have ended about 3.8 billion years ago [30, 31], or just under one billion years after the Earth was formed. It would have been difficult for complex life to develop and evolve under those conditions. However, the rain of meteorites might well have brought the chirally selected amino acids and other important organic molecules to Earth, since the Galaxy had been in operation nearly nine billion years before Earth was formed, so would have had plenty of time to create complex molecules and to establish their chirality.

Although one must be careful in science not to draw too many conclusions from coincidences, it is interesting none the less that primitive life forms are thought to have formed very shortly after the period of intense meteoritic activity ended [32, 33]. The first Earthly life forms are thought to have been Prokaryotes, which are so basic that they lack a cell nucleus. It is known that slightly more than one billion years later the signatures of living things were unmistakable, and photosynthesis was also clearly in evidence (http://en.wikipedia.org/wiki/Abiogenesis). Could all of this be the *result* of the meteoritic bombardment? Might it be that the cosmic stork is really just a dirty iceball—a comet? It is certainly not implausible to conclude that the intense meteorite bombardment of Earth brought the chirally selected amino acids and other important organic molecules to Earth, that they led to the development of early life forms, and that these evolved ultimately into the life forms that exist on Earth today. Of course, they might also have evolved into life forms on other planets, although not necessarily in the Solar system.

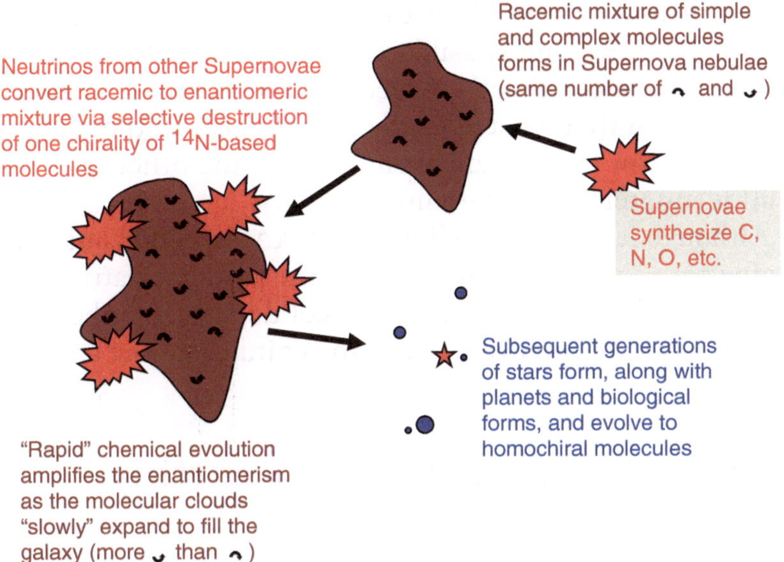
Racemic mixture of simple
and complex molecules
forms in Supernova nebulae
(same number of and)
Neutrinos from other Supernovae
convert racemic to enantiomeric
mixture via selective destruction
of one chirality of 14N-based
molecules
Supernovae
synthesize C,
N, O, etc.
Subsequent generations
of stars form, along with
planets and biological
forms, and evolve to
homochiral molecules
"Rapid" chemical evolution
amplifies the enantiomerism
as the molecular clouds
"slowly" expand to fill the
galaxy (more than)

FIGURE 8.2 Scenario by which molecular chirality is produced in the Galaxy in the context of the Supernova Neutrino Amino Acid Processing, or SNAAP, model. (From Boyd et al. [7]. Courtesy of International Journal of Molecular Sciences)

So let me again refer to the figure, which I introduced in Chap. 1, that illustrates how the SNAAP model works, now that you have the perspective of the intervening six chapters. I've reproduced that figure in Figure 8.2. Beginning in the upper-right corner, supernovae first produce the elements from which the molecules are made, and inject them into the interstellar medium. Interstellar chemistry creates the chiral molecules, either amino acids or their precursor molecules, in racemic mixtures, within the giant molecular clouds. Many supernovae occur in those clouds within their lifetime, each producing many electron antineutrinos with sufficient energy to convert ^{14}N to ^{14}C, and therefore to destroy that molecule. But it will be more difficult to destroy the molecules in which the antineutrino spin and the ^{14}N spin are aligned than those in which the spins are antialigned, and this, together with an asymmetry in the neutrino flux at the two poles of the neutron star, will produce a chiral selection. Chemical amplification in the grains, meteoroids, and comets enhances the level of enantiomerism, and Galactic effects mix the enantiomeric material throughout much of the Galaxy. Local compressions of the gas and grains, and possibly comets, will produce new stellar systems and planets. Finally, a second phase of amplification of the enantiomerism that existed in the components of the planets will lead to the amino acids, and eventually all the other molecules of life, all ultimately with the same chirality.

Although it might seem like the SNAAP model requires a lot of "moving parts" to make it work, all of the parts are naturally occurring in Nature. What the SNAAP model does is describe how they would all work together to produce the chirality of the amino acids.

8.5 Was the Development of Earthly Life a Fluke?

The evidence from the meteorites forces us to face the fact that the molecules of life, not only the amino acids but those molecules that are the precursors of our DNA, most likely have had their origin in outer space. While the biologically interesting molecules found on the Murchison meteorite were not DNA or RNA molecules themselves, it is generally felt that these molecules, which are far more complex than the amino acids, developed on Earth some time after the amino acids and nucleobases arrived. While the chemical pathways by which this evolution might have happened are not well defined, the very nature of evolution, that is, that the evolved species do not leave the traces of their evolutionary history, means that we can only guess the evolutionary pathways, and then test to see if they are plausible (see the article by Rode et al. [34] for a nice discussion of this). But it appears, as was discussed in Chap. 5, that the amino acids and nucleobases may well have been the crucial component in the development of life.

Does this mean that what happened on Earth could have happened lots of places in the cosmos? If the basic molecules of life are being made in outer space and distributed throughout the Galaxy, indeed throughout the Universe, it would appear that there is a good chance that we're not alone, and thus there may be creatures out there that have a lot in common with us. But a few words of caution are in order. Plaxco and Gross [35] discuss this in great detail, and conclude that although the creation of the basic molecules of life seems to be fairly easy, having them evolve into (we like to think) intelligent beings may be a bit trickier.

For example, we know that when large meteorites hit the Earth they can do more than just deliver biologically interesting molecules; they can also cause enormous damage, even causing the obliteration of most life for years if they are large enough. Earth was very fortunate in having Jupiter located in the Solar System where it is (as was discussed in Chap. 1). This enormous planet has basically swept the region around the inner planets clean of space debris (the natural type, not the pieces labeled "NASA", although the latter type is rapidly becoming a significant threat to space travelers) and prevented the debris lurking outside the region of the inner planets from getting to the inner planets. This may have spared Earth and the other planets inside Jupiter's orbit from nearly all of the devastating meteorite hits to which they would otherwise have been subjected. It would certainly have been crucial to the development of more and more sophisticated life forms to avoid occasional life extinguishing meteorite hits (although a few have occurred over time despite Jupiter's best efforts to eliminate that possibility).

In addition, Earth is located at just the right distance from the Sun for it to have liquid water. Is that essential for any form of life? It may be; scientists have tried to envision other life forms that might have evolved in the absence of water, but the chemical properties of water make it pretty ideal for allowing the development of life [35, 36]. Finally, although we can do experiments that suggest the chemical pathways by which more complex molecules are formed, it is not yet clear, for example, how natural evolutionary processes created cells with cell walls, which are certainly essential to all the complex life forms we know [35]. Was this a completely random occurrence?

References

1. J. Beacom, R.N. Boyd, and A. Mezzacappa, Black Hole Formation in Core-Collapse Supernovae and Time-of-Flight Measurements of the Neutrino Masses, Phys. Rev. D 63, 073011 (2001)
2. B.G. Elmegreen, Star Formation in a Crossing Time, Astrophys. J. 530, 277 (2000)

3. L. Hartmann, J. Ballesteros-Paredes, and E.A. Bergin, Rapid Formation of Molecular Clouds and Stars in the Solar Neighborhood, Astrophys. J. 562, 852 (2001)
4. C. Lunardini, Diffuse Neutrino Flux from Failed Supernovae, Phys. Rev. Letters 102, 231101–1 (2009)
5. R. Duncan and C. Thompson, Formation of Very Strongly Magnetized Neutron Stars—A.J. Drake, S.G. Djorgovski, J.L. Prieto, A. Mahabal, D. Balam, R. Williams, M.J. Graham, M. Catalan, E. Beshore, and S. Larson, Discovery of the Extremely Energetic Supernova 2008fz, Astrophys. J. 718, L127 (2010)
6. K.M. Ferriere, The Interstellar Environment of our Galaxy, Rev. Mod. Phys. 73, 1031 (2001)
7. R.N. Boyd, T. Kajino, and T. Onaka, Supernovae and the Chirality of the Amino Acids, Astrobiology 10, 561 (2010)
8. R.N. Boyd, T. Kajino, and T. Onaka, Stardust, Supernovae, and the Chirality of the Amino Acids, Int. J. Mod. Sci. 12, 3432 (2011)
9. V.I. Gol'danskii and V.V. Kuz'min, Spontaneous Breaking of Mirror Symmetry in Nature and the Origin of Life, Sov. Phys. Usp. 32, 1 (1989)
10. R.T. Garrod, S.L.W. Weaver, and E. Herbst, Complex Chemistry in Star-Forming Regions: An Expanded Gas-Grain Warm-up Chemical Model, Astrophys. J. 682, 283 (2008)
11. T.I. Hasegawa, E. Herbst, and C.M Leung, Models of Gas-Grain Chemistry in Dense Interstellar Clouds with Complex Organic Molecules, Astrophys. J. Suppl. 82, 167 (1992)
12. J.D. Laas, R.T. Garrod, E. Herbst, and S.L.W. Weaver, Contributions from Grain Surface and Gas Phase Chemistry to the Formation of Methyl Formate and its Structural Isomers. Astrophys. J. 728, 71-1 (2011)
13. D.K. Kondepudi and G.W. Nelson, Weak Neutral Currents and the Origin of Biomolecular Chirality, Nature 314, 438 (1985)
14. J.M. Pittard, *Mass Loaded Flows,* Diffuse *Matter from Star Forming Regions to Active Galaxies—A volume Honouring* John Dyson, ed. By T.W. Hartquist, J.M. Pittard, and S.A.E.G. Falle, Springer Dordrecht (2007); ibid, http://adsabs.harvard.edu/abs/2007dmsf.book..245P
15. V.V. Dwarkadas, Turbulence in Wind-Blown Bubbles Around Massive Stars, Phys. Scr. T132, 014024 (2008)
16. T.W. Hartquist, J.E. Dyson, and R.J.R. Williams, Mass Injection Rates Due to Supernovae and Cloud Evaporation in Starburst Superwinds, Astrophys. J. 482, 182 (1997)
17. C.L. Bennett, M. Halpern, G. Hinshaw, N. Jarosik, A. Kogut, M. Limon, S.S. Meyer, L. Page, D.N. Spergel, G.S. Tucker, E. Wollack, E.L. Wright, C. Barnes, M.R. Greason, R.S. Hill, E. Komatsu, M.R. Nolta, N. Odegard, H.V. Peiris, L. Verde, and J.L. Weiland, First-Year

WILKINSON MICROWAVE ANISOTROPY PROBE (WMAP) Observations: Preliminary Maps and Basic Results, Astrophys. J. Suppl. Series 148, 1 (2003)

18. N. Jarosik, C.L. Bennett, J. Dunkley, B. Gold, M.R. Greason, M. Halpern, R.S. Hill, G. Hinshaw, A. Kogut, E. Komatsu, D. Larson, M. Limon, S.S. Meyer, M.R. Nolta, N. Odegard, L. Page, K.M. Smith, D.N. Spergel, G.S. Tucker, J.L. Weiland, E. Wollack, and E.L. Wright, Seven-Year *WILKINSON MICROWAVE ANISOTROPY PROBE* (*WMAP*) Observations: Sky Maps,Systematic Errors, and Basic Results, Astrophys. J. Suppl. Series 192, 14 (2011)
19. S.R. Kulkarni, L. Blitz, and C. Heiles, Atomic Hydrogen in the Outer Milky Way, Astrophys. J. Lett. 259, L63 (1982)
20. K. Soai, T. Shibata, H. Morioka, and K. Choji, Asymmetric Autocatalysis and Amplification of Enantiomeric Excess of a Chiral Molecule, Nature 378, 767 (1995)
21. K. Soai and I. Sato, Asymmetric Autocatalysis and its Application to Chiral Discrimination, Chirality 14, 548 (2002)
22. S.P. Mathew, H. Iwamura, and D.G. Blackmond, Amplification of Enantiomeric Excess in a Proline-Mediated Reaction, Angew. Chem. Int. Ed. 43, 3317 (2004)
23. R. Breslow and M.S. Levine, Amplification of Enantiomeric Concentrations Under Credible Prebiotic Conditions, Proc. National Acad. Sciences 103, 12979 (2006)
24. A. Fujiwara, j. Kawaguchi, D.K. Yeomans, M. Abe, T. Mukai, T. Okada, J. Saito, H. Yano, M. Yoshikawa, D.J. Scheeres, O. Barnouin-Jha, A.F. Cheng, H. Demura, R.W. Gaskell, N. Hirata, H. Ikeda, T. Kominato, H. Miyamoto, A.M. Nakamura, R. Nakamura, S. Sasaki, and K. Uesugi, The Rubble-pile Asteroid Itokawa as Observed by Hayabusa, Science 312, 1330 (2006)
25. J. Saito, H. Miyamoto, R. Nakamujra, M. Ishiguro, T. Michikami, A.M. Nakamura, H. Demura, S. Sasaki, N. Hirata, C. Honda, A. Yamamoto, Y. Yokota, T. Fuse, F. Yoshida, D.J. Tholen, R.W. Gaskell, T. Hashimoto, T. Kubota, Y. Higuchi, T. Nakamura, P. Smith, K. Niraoka, T. Honda, S. Kobayashi, M. Furuya, N. Matsumoto, E. Nemoto, A. Yukishita, K. Kitazato, B. Dermawan, A. Sogame, J. Terazono, C. Shinohara, and H. Akiyama, Detailed Images of Asteroid 25143 Itokawa from Hayabusa, Science 312, 1341 (2006)
26. H. Yano, T. Kubota, H. Miyamoto, T. Okada, D. Scheeres, Y. Takagi, K. Yoshida, M. Abe, S. Abe, O. Barnouin-Jha, A. Fujiwara, S. Hasegawa, T. Hashimoto, M. Ishiguro, M. Kato, J. Kawaguchi, T. Mukai, J. Saito, S. Sasaki, and M. Yoshikawa, Touchdown of the Hayabush Spacecraft at the Muses Sea on Itokawa, Science 312, 1350 (2006)

27. W.H.-P. Thiemann and U. Meierhenrich, ESA Mission ROSETTA Will Probe for Chirality of Cometary Amino Acids, Orig. Life Evol. Biosphere 31, 199 (2001)
28. P. deMarcellus, C. Meinert, M. Nuevo, J.-J. Filippi, G. Danger, D. Deboffle, L. Nahon, L.L.S. d'Hendecourt, and U.J. Meierhenrich, Non-racemic amino acide production by ultraviolet irradiation of achiral interstellar ice analogs with circularly polarized light. Astrophys. J. 727, L1 (2011)
29. Yu. Efremenko and W.R. Hix, Opportunities for Neutrino Physics at the Spallation Neutron Source (SNS), J. Phys: Conf. Series 173 012006 (2009)
30. R.G. Strom, R. Malhotra, T. Ito, F. Yoshida, and D.A. Kring, The Origin of Planetary Impactors in the Inner Solar System, Science 309, 1847 (2005)
31. R. Gomes, H.F. Levison, K. Tsiganis, and A. Morbidelli, Origin of the Cataclysmic Late Heavy Bombardment Period of the Terrestrial Planets, Nature 435, 466 (2005)
32. S.J. Mojzsis, G. Arrhenius, K.D. McKeegan, T.M. Harrison, A.P. Nutman, and C.R.L. Friend, Evidence for Life on Earth Before 3800 Million Years ago, Nature 384, 55 (1996)
33. J.W. Schopf, A.B. Kudryavtsev, D.G. Agresti, T.J. Wdowiak, and A.D. Czaja, Laser-Raman Imagery of Earth's Earliest Fossils, Nature 416, 73 (2002)
34. B.M. Rode, D. Fitz, and T. Jakschitz, The First Steps of Chemical Evolution Towards the Origin of Life, Chemistry and Biodiversity 4, 2674 (2007)
35. K.W. Plaxco and M. Gross, *Astrobiology*, Johns Hopkins University Press, Baltimore, 2006
36. L.J. Rothschild and R.L. Mancinelli, Life in Extreme Environments. Nature 409, 1092 (2001)

9. What About the Rest of the Universe?

Abstract What conditions would be required elsewhere in the Universe to support life? Perhaps we can explore this possibility by studying the conditions in which Earthly extremophiles exist. We find that many living forms can exist under vastly different circumstances than what we would ordinarily think would be required to support life. This chapter describes some of them, then extends these conditions to outer space to infer the possible existence of extraterrestrial life. Finally, the several decade Search for Extraterrestrial Intelligence, SETI, is described.

9.1 Are We Alone?

In any discussion of the origin of life, one immediately faces the question about the ubiquity of the solutions one devises, that is, if we have figured out how life originated on Earth, is there any reason to assume that Earth is so special that life could not have originated in other places in the Universe as well? Although there seem to be preconditions to the existence of life anywhere, it would surely be surprising if the conditions that made it possible for life to evolve to its present state on Earth didn't exist elsewhere in the Universe. Indeed, the conditions of Earth may not even be as strict as one might imagine, which is immediately realized by studying the extreme versions of life that are found on Earth. Let's look at some of those.

R.N. Boyd, *Stardust, Supernovae and the Molecules of Life: Might We All Be Aliens?*, Astronomers' Universe, DOI 10.1007/978-1-4614-1332-5_9,

9.2 Extremophiles on Earth

Of course, it is possible that very different life forms evolved from the same basic chemicals of which we are made. Indeed, it may be difficult to even recognize the critters that inhabit other planets, or even obscure places on Earth. This gets us back to our discussion of the definition of life. While we may not want to wait around for our candidate life forms to multiply, it would certainly be essential for them to metabolize their food, whatever that might be. So that might be the most plausible means of detecting that our sample contains something that is alive. But even this might take on forms with which we are not familiar; even on Earth critters have been found that eat things that would literally poison the life forms we know and love. For example, extremophiles have been found that happily consume sulphur [1] and possibly even arsenic [2], although the latter result is not without challenge [3]. *Ferroplasma acidarmanus* has been found to grow at a pH of 0.0 in acid mine drainage [4] from Iron Mountain in California. There it apparently thrives on a mixture of sulphuric acid that is laced with high levels of copper, arsenic, cadmium, and zinc. In the case of the sulphur eating bacteria, these were suggested as being capable of cleaning up the sulphuric acid that results from mine tailings. Thus some of our more extreme "relatives" with which Nature has presented us may be better at maintaining our natural resources than the owners of "intelligent life." In any event, the search for life must cast a wide net for possibilities that might not look very much like the living things we could identify easily.

However, we know of living things that are more than just those with dietary oddities. So called extremophiles are known to push the limits of temperature, acidity, drought, and probably a lot of other things well beyond what humans could tolerate. To quote Penelope Boston [5], "When we look at our own planet's most challenging environments, we are really looking for clues to what may be the normal conditions on other planets. We want a hint of what we may be searching for when we investigate those other worlds for signs of life." So our searches of our Earthly environment for the weirdest critters may provide helpful signals for our searches for life beyond Earth. Boston's subtitle for her article is "What is extreme here may be just business-as-usual elsewhere."

Let me discuss a few examples of extremophiles in enough detail to give the reader some idea of what "extreme" really means. There are hydrothermal vents in the deep sea (Wikipedia: "Hydrothermal vent") that spew forth material: these are called either white smokers or black smokers. The white smoker at the Champagne vent, northwest Eifuku volcano, Marianas Trench Marine National Monument, produces liquid carbon dioxide. Recall that in the deep ocean the pressure can be several hundred times higher than it is at the Earth's surface; this is the reason that the carbon dioxide can even exist as a liquid (it is either a gas or a solid at atmospheric pressure). Black smokers are little ecosystems in themselves. They spew forth a variety of chemicals which build up along their sides to produce a sort of chimney. Although the temperature of sea water near these black smokers is about 2°C, just above the freezing point of water at the Earth's surface, the temperature of the material coming from the vent can be as high as several hundred degrees C, or several times the boiling temperature of water at Earth's surface. Of course, the water doesn't boil at that ocean depth, again because of the high pressure.

Despite their high temperature, the black smokers have been found to be home to a variety of living entities. In particular, Kashefi and Loveley [6] analyzed mircoorganisms from the black smoker "Finn," which is located in the Mothra hydrothermal vent field (47°55.46′N and 129°06.51′W) along the Endeavor segment of the Juan de Fuca Ridge in the northeast Pacific Ocean. One of the organisms they studied was designated "strain 121." It was isolated at 100°C, the boiling point of water at atmospheric pressure. Some iron oxide was included in the mixture. Cell growth and iron reduction were monitored. Strain 121 is typical of archaea, single celled organisms, the cells of which do not contain nuclei (see discussion in Chap. 1). It grew at temperatures between 85°C and 121°C, so continued to exist even at that extreme temperature. Because of the ability of this and other "hyperthermophiles" (organisms that can live at temperatures in excess of 80°C) to withstand high temperatures, they have been suggested as a possible early life forms for Earth, which also existed at an extremely high temperature in its stage in which life began (http://www.theguardians.com/Microbiology/gm_mbm04.htm). This point is supported by Davies [7], who strongly advocates the microbial life that exists near the black smokers as possibly being closely related to the first life forms that existed on Earth.

The heat-loving critters were actually first discovered in Yellowstone National Park, although the Yellowstone extremophiles are not quite as extreme as the black smokers. This is described nicely by Rothschild and Mancinelli [8] and by Sarah Bordenstein (http://serc.carleton.edu/microbelife/topics/octopusspring/index.html) for one of the hotsprings in Yellowstone National Park, the "Octopus Spring," which is highly alkaline (pH of 8.8–8.3). It is located about eight miles north of Old Faithful geyser. As described by Bordenstein, the spring has a primary source from which the water flows at 95°C. Its drainage channels radiate like the arms of an octopus (well, sort of, anyway), hence the name of the spring. As the water flows to one part of the accompanying pond, its temperature cools to about 83°C, but then increases to 88°C when a new surge of hot water is emitted. In this environment pink filamentous communities thrive, at a distance of about 2 m from the crystal blue pool. Elsewhere in the pond the water cools to less than 75°C, a temperature that permits growth of microbial mat communities that include cyanobacteria, green sulfur bacteria, and green non-sulfur bacteria. There one can see the growth of a thermophillic cynobacterium in this part of the pond. At 65°C a more complex microbial mat forms one bacterium on the top overlaying other bacteria, including species of a photosynthetic bacterium. Some superb photographs that serve to illustrate the situation better than words can have been taken of Octopus Spring by Alan Treiman; one of these is shown in Figure 9.1, with some additional explanation given in the caption.

At the other extreme of temperature are the cold-loving entities, the "cryophiles." These include cyanobacteria, bacteria, fungi, spores, etc. (see Scientist's Notebook, http://spacescience.spaceref.com/newhome/headlines/ice.pdf). These were discovered in Lake Vostok, in Antarctica by Richard Hoover of NASA's Marshall Space Sciences Laboratory and S.S. Abyzov of the Russian Academy of Sciences. The organisms found are thought to have been trapped in the ice about 400,000 years ago. They exist in a dormant state, but are still metabolizing at an extremely low level. Their existence might bode well for finding life forms on cold planets or moons of planets, for example, on Europa, one of Jupiter's moons.

However, the most significant aspect of life existing at cryogenic temperatures may be, as noted in Chap. 5, that the amino

FIGURE 9.1 Octopus Springs. The hot water, greater than 90°C, comes up in the small pool to the right of the main, blue pool. Part of the water leaves immediately in the stream to the right, where the pink streamers are located. The rest spills over in to the large blue pool, and then flows out to the upper right where the microbial mats were found. (Courtesy of Allan Treimann, Lunar and Planetary Institute)

acids and nucleobases can be assembled, even at temperatures so low that ordinary life, at least human life, could not be sustained. However, if life forms tried to exist at such temperatures they would need to be very careful; when water freezes inside a cell it almost always ruptures the cell when it expands from its liquid state to its ice state. But the molecules of life seem to have no difficulty assembling at such temperatures; indeed the assembly of at least some of those molecules appears to require those temperatures. Of course, this may have important ramifications on our conclusions about the source of the molecules of life. It might be difficult to imagine cold conditions on the early Earth, although the day-night cycles might have been helpful in that regard, and such cycles might even be important for producing some of the important molecules. Of course, the low temperatures come naturally with meteoroids and comets for much of their lives, so the ultimate conclusion from the low temperature molecule production [9] it must have occurred on comets. And, as noted in Chap. 6, the molecules produced in the insides of such objects would not be subject to the processing by ultraviolet radiation. But they would be sensitive to strong external magnetic fields. More importantly, they would also be sensitive to neutrino processing! This is an important aspect of the SNAAP model [10, 11].

The bacterium *D. Radiodurans* has been found to be able to withstand both gamma radiation and ultraviolet radiation in extremely high doses. However, its ability to do so apparently revolves around its ability to repair the damage to its DNA that results from the high radiation doses. This involves reassembling the fragmented DNA [12, 13].

Add to these the critters that prefer very low or very high pH levels, that is, very acidic or basic environments, those that exist in very salty or very metal rich environments, those that exist at very high pressure, and others (http://en.wikipedia.org/wiki/Extremophile), and it becomes clear that the possibilities for life are enormous indeed, and may lie far from what humans would regard as "normal."

The obvious question to ask, however, is how do extremophiles exist in such extreme surroundings? Rothschild and Mancinelli [8] deal with this question for several extremophiles. The thing that seems to work best is for the extremophiles to simply keep the external environment out. For example, *Cyanidium caldarium* and *Dunaliella acidophila* are found at an extremely acidic pH of 0.5. But Pick [14] and Beardall and Entwisle [15] found them to have a nearly neutral cytoplasm (which is the thick liquid that holds all of the internal entities of a cell except the nucleus). This can only be the case if the internal part of their cells is buffered against the highly acidic external environment. Another possibility is to remove the extreme condition as rapidly as possible. Here the example is heavy metal-resistant bacteria. Niles [16] found that they use an efflux pump to remove zinc, copper, and cobalt, but not mercury, which is volatilized.

If it is impossible to keep the environment out, then the extremophiles adopt protective mechanisms, alter their physiology or have specific repair mechanisms. Of particular note in this regard is the nucleic acids, for which function and structure are closely linked. DNA is especially vulnerable to high temperature, radiation, oxidative damage, and dehydration. In some cases, the extremophiles have adopted multiple ways to solve the problem of living in a particularly hostile environment.

But it would not be appropriate to leave the extraordinary without mentioning the striking features of the very ordinary, a pond dwelling paramecium. *Paramedium bursaria* appears green under a microscope because each cell hosts hundreds of chlorella algae that supply it with sugar and oxygen. It can reproduce asexually, splitting into two identical daughter cells, but occasionally docks with another cell to exchange small capsules that hold DNA in order to correct damaged DNA. They can swim ten times the length of their body in 1 second. Furthermore, they have almost 40,000 genes, about double the number in a human cell [17]. Even "ordinary" can be pretty amazing!

9.3 And From Outer Space?

Of course, identifying extremophiles and associating their environmental preferences with the conditions of early Earth certainly does not guarantee that they really are early life forms. This identification would require that we could reconstruct all the intermediate stages of life, a very tall order at present, and possibly for a long time to come. Most discouragingly, the best we could ever do would be to reconstruct a plausible succession of living forms, but even that would not prove that these were the actual forms.

In recognition of this, and in an attempt to circumvent the successive steps questions, a group set out in 1960 to simply detect the presence of extraterrestrial beings. This effort was called the Search for Extraterrestrial Intelligence, or SETI, and its founding father was Frank Drake (of the Drake equation, discussed in Chap. 4). Later Carl Sagan played an important role, both scientific and political, in furthering the SETI program. The main question they had to address was: how do you "look" for extraterrestrial beings? You certainly do not just go somewhere beyond Earth and look for them (with the exceptions of our Moon and Mars); you need to try to identify plausible places for life, and search there. But space is a big place, and travel times to plausible places can be many human lifetimes, so one has to figure out what would constitute an identification of an extraterrestrial "life signal" that would be convincing even in the absence of an actual sighting. What was settled on was a detection of a radio signal from outer space that was at some frequency that would be of particular significance to life. It should be noted that the use of radio signals for detecting the presence of extraterrestrial life had been suggested even just before the beginning of the twentieth century by Nicola Tesla, and later by such notables as Guglielomo Marconi and Lord Kelvin. And a paper published in 1959 by Cocconi and Morrison suggested searching in the microwave part of the electromagnetic spectrum for life signals, setting the stage for the Drake experiment in 1960. Although there were many frequencies suggested as "life signals," technology caught up with politics (usually it's the other way around, if it happens at all) when it became possible for radio telescopes to monitor a huge number of radio frequencies simultaneously. This obviated the need to have our life signal frequency match that of the civilizations for which we were searching.

The advantage of using radio signals to search for life signals is fairly obvious to astronomers: radio telescopes (see Chap. 4) can be huge, can be made in multiple dishes so as to present a huge photon collecting area, and can have good capability for localizing the source of the signals. They also are not absorbed by the Earth's atmosphere, rendering Earth-based observations adequate for the task, and circumventing the need to put large objects into space.

Unfortunately (or, perhaps, fortunately) no convincing signals have yet been detected. This does not mean that there are no intelligent (by our standards, anyway) beings out there. They might be too far away for us to detect their life signals, or they may not have yet achieved the capability to send strong radio signals into space, or they may have advanced in their communications beyond the use of radio signals, or they may not be especially interested in trying to connect with what they would identify as aliens—that would be us: all those factors that we discussed in the context of the Drake equation. Searching for extraterrestrials can be a tough business! And, in very difficult fiscal times, even though SETI has been mostly privately funded, obtaining funding to continue such efforts can be just as difficult.

In any event, the lack of convincing positive results in our searches for extraterrestrials leads to what is known as the Fermi paradox, suggested by Enrico Fermi in the 1950s (http://en.wikipedia.org/wiki/SETI):

> The size and age of the universe incline us to believe that many technologically advanced civilizations must exist. However, this belief seems logically inconsistent with our lack of observational evidence to support it. Either (1) the initial assumption is incorrect and technologically advanced intelligent life is much rarer than we believe, or (2) our current observations are incomplete and we simply have not detected them yet, or (3) our search methodologies are flawed and we are not searching for the correct indicators.

This triad of possible conclusions would seem to cover all possibilities!

So perhaps we have been looking in the wrong places or with the wrong tools for extraterrestrial life. With the advent of the results from the Kepler Mission, more and more candidate planets are becoming known. Surely some of those will be "Goldilocks planets," that is, not too hot, not too cold. And with modern technology scientists can search for different signatures of life than intercepted radio signals, such as production of methane and other gases that could be detected in the planetary atmospheres. Indeed such searches are moving to the fore [18].

Then there is the possibility that Francis Crick might be partially correct in asserting that life was brought to Earth by an intelligent civilization from another planet—the Panspermia hypothesis. As noted by Davies [19], the visitors might even have left their mark in a way that we would find difficult to detect with our present technology. For example, they might have implanted some specific DNA modification on whatever form of DNA they found when they visited. We must take into account the very real possibility (probability?) that there exist technically more advanced civilizations in our Galaxy than the one with which we are familiar!

Although it seems obvious that we should seek communication with extraterrestrial beings, simply on the basis that more information is better than less information, and besides, the existence thereof is an extremely intellectually interesting question. However, there have been voices raised in opposition to such communication. Perhaps the most notable is Stephen Hawking, who is quoted (http://en.wikipedia.org/wiki/Stephen_Hawking) as saying: "To my mathematical brain, the numbers alone make thinking about aliens perfectly rational. The real challenge is to work out what aliens might actually be like." He goes on to warn against contacting aliens, but rather doing our best to avoid them. His thinking in this regard: "If aliens visit us, the outcome would be much as when Columbus landed in America, which didn't turn out well for the Native Americans." Perhaps it would be advisable to keep a low cosmic profile!

Let's give a bit more thought to Davies' suggestion of the possibility of an alien DNA modification. As Davies [19] and many others note, there is a lot of our DNA that doesn't seem to be involved in any essential aspect of our lives—it is generally referred to as "junk" sections of our DNA. Might our hypothetical advanced aliens figured out that they could store a "Hi Earthling" message in our DNA that wouldn't affect the function of the DNA? And perhaps someday we will have become clever enough to read the message. Or perhaps they were more insidious, and reprogrammed our DNA so that when they return to Earth they will flip a genetic switch and we will suddenly all change to a slave mentality. The possibilities are limitless.

None the less, the basic chemicals of life are made in outer space. The meteoritic evidence is strong that the basic molecules of life—both the amino acids and the nucleobases—are made in outer space. Furthermore, the fact that the amino acid chirality is established in space would suggest that the molecules that are produced in space are the triggers of terrestrial life. Of course, other planets could also not have avoided being hit by many such meteorites, so all planets would have been seeded with these molecules. And, as discussed in Chap. 4, perhaps other planetary systems also contain Earth-like planets at the right distance from their stars to allow them to have liquid water, and perhaps even Jupiters to sweep out the space detritus inside its orbit, and thus to protect the inner planets. There are, after all, more than 100 billion stars in our Galaxy alone, and there should be more than one of them that would satisfy the special conditions that Earth enjoys. That being the case, it is only natural to believe that we share a universe with other brethren. Although it is very unlikely that they would have evolved in the same way we did, since there is a lot of chance involved, even in just the modification of DNA by cosmic ray induced mutations, there is none the less a large probability that *something* evolved on other planets. The X-lings may look quite different from us; they may be green and have antennae, but their basic molecules of life may be pretty similar to ours, and their cells might be pretty similar too. With a slightly better twist of molecular evolutionary fate we might also have been green and have had antennae!

But, like it or not, we may have quasi brothers and quasi sisters in the far distant reaches of the Universe, and maybe even much closer than that!

References

1. K.L.I. Norlund, G. Southam, T. Tyliszcak, Y. Hu, C. Karunakaran, M. Obst, A.P. Hitchcock, and L.A. Warren, A Novel Syntrophic Microbial Sulphur Metabolising Consortia: New Evidence for Microbial Global Carbon Cycling, Env. Sci. Tech. 43, 8781 (2009)
2. F. Wolfe-Simon, J.S. Blum, T.R. Kulp, G.W. Gordon, S.E. Hoeft, J. Pett-Ridge, J.F. Stolz, S.M. Webb, P.K. Weber, P.C.W. Davies, A.D. Anbar, and R.W. Oremland, A Bacterium that Can Grow by Using Arsenic Instead of Phosphorus, Science Express, downloaded from Sciencemag.org, Dec. 2 (2010)
3. E. Pennisi, Concerns About Arsenic-Laden Bacterium Aired, Science 332, 1136 (2011)
4. K.J. Edwards, P.L. Bond, T.M. Gihring, and J.F. Banfield, An Archael, Iron, Oxidizing, Extreme, Acidophile, Important, Acid, Mine, Drainage. Science 287, 1796 (2000)
5. P. Boston, The Search for Extremophiles on Earth and Beyond, What is Extreme Here May Be Just Business-as-Usual Elsewhere, Ad Astra Astrobiology Issue Expanded Edition, Ad Astra Magazine, The Astrobiology Web (2011). Courtesy of Ad Astra Magazine and P. Boston
6. K. Kashefi and D.R. Lovely, Extending the Upper Temperature Limit for Life. Science 301, 934 (2003)
7. P. Davies, *The 5th Miracle: The Search for the Origin and the Meaning of Life*, Simon and Schuster, NY, NY, 1999
8. L.J. Rothschild and R.L. Mancinelli, Life in Extreme Environments. Nature 409, 1092 (2001)

9. M. Levy, S.L. Miller, K. Brinton, and J.L. Bada, Prebiotic Synthesis of Adenine and Amino Acids Under Europa-Like Conditions, Icarus 145, 609 (2000)
10. R.N. Boyd, T. Kajino, and T. Onaka, Stardust, Supernovae, and the Chirality of the Amino Acids, Int. J. Mod. Sci. 12, 3432 (2011)
11. R.N. Boyd, T. Kajino, and T. Onaka, Supernovae and the Chirality of the Amino Acids, Astrobiology 10, 561 (2010)
12. J.R. Battista, Against All, Odds, Survival, Strategies, Radiodurans. Annu. Rev. Microbiol. 51, 203 (1997)
13. J.R. Battista, in DNA Damage and Repair, Vol I: DNA Repair in Prokaryotes and Lower Eukaryotes. eds. J.A. Nickoloff and M.F. Hoekstra, (Humana, Totowa, NJ), p. 287 (1998)
14. U. Pick, in *Enigmatic Microorganisms and Life in Extreme Environments* (ed. J. Seckbach) 467 (Kluwer, Dordrecht) 1999
15. J. Beardall and L. Entwisle, Internal pH of the Obligate, Acidophile *Cyanidium caldarium* Geitler (Rhodophyla?). Phycologia 23, 397 (1984)
16. D.H. Niles, Heavy Metal, Resistant, Bacteria, Extremophiles, Molecular, Physiology, Biotechnological, Use of *Ralstonia* sp. CH34. Extremophiles 4, 77 (2000)
17. M. Grunbaum, Paramecium, Discover, May, p. 16 (2011)
18. Y. Bhattacharjee, A Distant Glimpse of Alien Life? Science 333, 930 (2011) DOI: 10. 1126/science.333.6045.930
19. P. Davies, *The Eerie Silence: Renewing Our Search for Alien Intelligence*, Houghton, Miflin, Harcourt, NY, 2010

Glossary

Absorption spectra Spectra resulting from the absorption in a star's surface of the radiation produced inside the star, and which backlights the surface region.
Adaptive optics This refers to the ability of a telescope to shape its primary mirror on a short time scale in order to minimize the distortion caused by the Earth's atmosphere, and optimize its resolution.
Adenine One of the mucleotide bases of which DNA is comprised.
Algae Unicellular or multicellular chlorophyll containing plants occurring in fresh or salt water.
Alpha-amino acids One class of amino acids that is distinguished from other amino acids by the location of its different groups. Alpha-amino acids are the ones that are important to Earthly life.
Amino acid These are molecules that are characterized by all having an amino group, NH2, and a carboxyl group COOH (see Chap. 4 for a picture), along with the other atoms that distinguish them from each other. The amino acids are components of the proteins on which we rely for life.
Amino group This is NH_2, one of the basic components of all amino acids along with the carboxyl group.
Amplification This refers to the process by which tiny levels of enantiomerism can be increased to much larger levels, or even to homochirality. This could occur either on a planet or in the cosmos, although the mechanisms in the two environments would be quite different. This could also refer to increasing the total number of amino acids.
Angular momentum A conserved quantity in physics that is associated with rotational motion.
Archaea Single celled organisms that have no nucleus, and constitute one of the three basic groups of archaea, bacteria, or eukaryotes.

R.N. Boyd, *Stardust, Supernovae and the Molecules of Life: Might We All Be Aliens?*, Astronomers' Universe, DOI 10.1007/978-1-4614-1332-5,

Asteroid Belt A huge system of interplanetary debris that exists between the orbits of Mars and Jupiter.
Astronomical unit Abbreviated "AU", this is the mean distance between the Earth and the Sun, which is 1.5×10^{13} cm.
Asymptotic giant branch stars Stars that have completed burning the hydrogen in their cores, and have evolved from the characteristics of that burning mode into those which characterizes helium burning.
Autocatalysis This refers to chemical replication schemes by which molecules replicate themselves.
Bacteria Single celled organisms that, like archaea, have no nucleus, and constitute one of the three basic groups of archaea, bacteria, or eukaryote. However they also differ from archaea in several ways.
Baryon Particles that are made of three quarks. Protons and neutrons are baryons; photons, electrons, and neutrinos are not. Pions also are not.
Baryon conservation One of the laws of physics that says that the number of baryons in any reaction or decay must be conserved.
Beta decay The process by which an unstable nucleus emits either an electron and an electron antineutrino, or emits a positron and an electron neutrino, or captures an electron and emits an electron neutrino. In so doing it will form a more stable nucleus.
Big bang The creation event of our Universe.
Binding energy (of nuclei) The energy by which a nucleus or molecule is bound, that is, the energy that would be required to break it apart into at least some of its constituents.
Black hole The final state of a very massive star after it has completed all its stages of stellar evolution. No matter can escape from a black hole. Massive black holes also exist at the centers of galaxies.
Black smokers Hydrothermal vents in the ocean floor, from which pour a variety of substances.
Bremsstrahlung Literally, "braking radiation," which is produced when a charged particle is accelerated or decelerated.
Buckingham effect The result of the interaction between a nucleus that has a nonzero spin and an external magnetic field. This was originally devised to describe how nuclear magnetic resonance could be used to differentiate between molecules of opposite chirality.

Cambrian explosion The event in Earth's history 500–600 million years ago when there was an incredible explosion in the diversity of life forms.

Carbonaceous chondrites These are one kind of meteorite, which appears particularly capable of withstanding the high temperatures associated with passage through the Earth's atmosphere. These have been found to contain amino acids.

Carboxyl group This is COOH, one of the basic components of all amino acids along with the amino group.

Carbon burning One of the stages of stellar evolution, following helium burning and preceding neon burning.

Catalysis In catalysis, the "catalyst" nucleus or molecule enables a process in which new nuclei or molecules are formed, and the original nucleus or molecule is returned.

Centigrade One of several temperature scales. On this scale, water freezes at 0°C and boils at 100°C.

Cepheid variables Stars that have an oscillatory output, the frequency of the oscillations of which is related to the absolute luminosity of the star. Cepheids, therefore, have been used as distance indicators.

Chandra X-Ray observatory This is one of the space borne observatories designed to detect x-rays. This observatory was named in honor of W. Chandrasekar.

Charged-current weak interactions Those weak interactions that change one nucleus to another one of nearly equal mass, but with different numbers of protons and neutrons.

Chart of the nuclides The chart of all the known nuclides, stable and unstable. This includes all known isotopes of every element.

Chemical evolution (of the Galaxy) The buildup in time of the elements in the interstellar medium.

Chemical evolution (of biomolecules) This describes the evolution of the biomolecules from simple amino acids to peptides, or the nucleotides, to the more complex DNA and RNA.

Chirality Handedness. Chiral molecules have an "opposite" that has the opposite chirality. Chirally opposite molecules cannot be translated in such a way that they are identical, but do appear identical if one of them is "reflected in a mirror."

Chlorophyll The green coloring of matter that enables production of carbohydrates by photosynthesis.

Churyumov Gerasimenko A comet, which is the destination of the European Space Agency mission ROSETTA.
Circularly polarized light Light for which the electric field vector appears to rotate (see figure in Chap. 4). In right circularly polarized light, the electric field vector appears to rotate in a clockwise direction when the light is viewed in the direction of travel, and vice versa for left circularly polarized light.
CNO cycle This describes the details of the dominant form of hydrogen burning, in which four hydrogen nuclei get fused into a helium nucleus, in stars more massive than several solar masses.
Cold dark matter One of the hypothetical forms of dark matter, in which the particles are assumed to exist at low temperature.
Comet A celestial object, often quite large (compared to a meteorite, but not to a planet) that is usually composed of ice and dust.
Conservation laws The laws of physics that govern processes and forms. Examples include conservation of energy and conservation of angular momentum.
Core-collapse supernovae Those supernovae that are produced by massive stars when they have completed all their stages of stellar evolution.
Cosmology The study of the origin of our Universe. This also includes identification of the observable signatures of our Universe's birth event.
Coulomb barrier The barrier resulting from the interaction of two charged particles of the same sign. In particles at low energy the Coulomb barrier works to prevent the particles from undergoing any interaction other than the electromagnetic interaction, although they can interact via the strong interaction, and undergo fusion, via quantum mechanical tunneling.
Cross section The reaction probability for one particle to interact with another. See reaction probability.
Cryophiles Cold-loving creatures, a class of extremophiles.
Cyanobacteria Blue-green bacteria or archaea; its cells have reached a relatively high level of sophistication, at least for prokaryotes.
Cyanidium caldarium A living organism that can tolerate extremely low pH values, that is, very acidic environments.
Cytoplasm The liquid that resides between the cell walls and the nucleus in eukaryota, and within the cells of non-eukaryota cells.

Cytosine One of the nucleotide bases of which DNA is comprised.
Dark energy The entity that appears to cause the expansion of the Universe to accelerate.
Dark Matter The form of matter, as yet unknown, that exerts a gravitational force on other types of matter, and so influences the motion of galaxies.
Degeneracy pressure The pressure created by the Pauli Principle, which states that only one identical particle of a particular type (for example, electrons or neutrons) can occupy a single quantum state. The degeneracy pressure due to electrons is the pressure that maintains the sizes of atoms, and of white dwarfs. That due to neutrons and protons maintains the sizes of atomic nuclei, and of neutron stars.
Deuteron The nucleus of heavy hydrogen, that is, a nucleus that contains one proton and one neutron.
DNA Deoxyribonucleic acid. This is one of the basic chemicals of life. It contains the instructions for assembling proteins within cells, and also contains all of the genetic information essential for replication.
Doppler shift The shift in the spectral lines of atoms or ions with motion. When the atoms or ions are moving away from the observer the characteristic lines will be "red shifted."
D. Radiodurans A living organism that can tolerate, via repair mechanisms, very high levels of radiation.
Drake equation The equation, of sorts, that categorizes the different factors that would be required for some extraterrestrial civilization to send radio signals that we would interpret as indicating the existence of an advanced civilization.
Dunaliella acidophia A living organism that can tolerate extremely low pH, that is, highly acidic, environments.
Dust grains The tiny (of order 10 μm) grains of molecules that form in the interstellar medium.
Efflux pump Mechanism by which cells rid themselves of toxic entities, for example, zinc, copper, or cobalt.
Electric field vector The vector that specifies both the strength and the direction of the electric field.
Electric field The distribution of that physical entity that affects charged particles and, together with the magnetic field, represents the photons as they move through space.

Electromagnetic radiation The radiation that is the result of electric and magnetic fields. The particles of electromagnetic radiation are called photons.

Electromagnetic force/interaction One of the four basic interactions of physics.

Element Atoms of a particular element are characterized by the number of protons in their nucleus. For a neutral atom (that is, if it is not ionized), the number of electrons the atom has will be equal to the number of protons. Hydrogen, helium, oxygen, carbon, tin, etc. are elements.

Elliptically polarized light Light that has its electric field vectors somewhat out of phase, but not necessarily the 90° out of phase that would produce circularly polarized light, or not necessarily with two equal strength components of the electric field.

Enantiomeric excess The amount by which one chirality exceeds the other, divided by their sum. It is usually also expressed as a percentage, so the difference divided by the sum is multiplied by 100.

Endothermic Refers to reactions that absorb energy, that is, that require energy to make them occur.

Enzyme Proteins that act as catalysts to, among other functions, increase the rates of metabolic reactions.

Erg A unit of energy. A 93 mile per hour fastball has 12 billion ergs of energy.

Electron A light fundamental particle. It has a spin angular momentum of $\hbar/2$ and a negative charge.

Elliptically polarized light Light that has out of phase components, compared to circularly polarized light, or has unequal amplitudes of the two electric field vectors.

Enantiomeric Referring to a medium in which the two possible chiral states are not equally populated.

Eukaryotes Organisms that are the most complex of the three groups: archaea, bacteria, and eukaryote. Specifically, eukaryotes have a nuclear envelope in which their genetic material is contained, whereas archaea and bacteria have no nucleus.

Europa One of the moons of Jupiter.

Event horizon The radius at which escape from a black hole becomes impossible.

Evolution The natural progression from one entity to another. Specifically, this describes how species change with time.

Excited (nuclear or atomic) state An allowed state in which the nucleus or atom can exist that is at a higher (less tightly bound) energy than the ground state.

Exothermic Refers to reactions that give off energy when they occur.

Extraterrestrial Not of the Earth.

Extremophiles Creatures that prefer living conditions very different from those human beings prefer. These might include temperature extremes, pressure extremes, unusually acidic or alkaline conditions, etc. They might also be able to withstand extremes that humans could not, for example, high radiation levels.

Fahrenheit One of several temperature scales. On this scale absolute zero is –459.67°F, water freezes at 32°F, and it boils at 212°F.

Faraday effect The result of putting molecules in a magnetic field. The effect appears as if a ring of charges is induced to circulate in the molecules.

Fermi paradox The statement, by Enrico Fermi, about the three possibilities by which we have not observed extraterrestrial life.

Fission Breaking apart of a heavy object into two or more lighter objects.

Flavor These describe the different types of neutrinos, that is, the flavors of neutrinos are electron, muon, and tau. This term also applies to the different types of quarks.

Fungi Life forms characterized by an absence of chlorophyll, for example, mushrooms, yeasts, and molds.

Fusion Combining two lighter constituents to make a heavier object.

Galaxy The collection of stars that constitute the neighbors of the Sun, and move in concordance with it. Our galaxy, the Milky Way Galaxy, contains 100–400 billion stars, and is 100,000 light years in size. The distance to the nearest adjacent galaxy, the Andromeda Galaxy, is 2.5 million light years.

Gamma rays Electromagnetic radiation, or photons, that are at the high energy end of the electromagnetic energy spectrum.

Gauss A measure of the strength of a magnetic field. The Earth's magnetic field has a strength of roughly one gauss.

Genes A unit of heredity in a living organism. Genes are subunits of DNA and RNA that, for example, code for a type of protein.

Global chirality Chirality that is consistently the same everywhere, that is, does not vary from one place to another.

Green sulfur bacteria Living organisms observed in the Octopus Spring in Yellowstone National Park.

Green non-sulfur bacteria Living organisms observed in the Octopus Spring in Yellowstone National Park.

Ground state The lowest energy state in which an entity—an atom or a nucleus—can exist.

Gravitational potential energy The energy associated with position. For example, when you hold an object in the air it has gravitational potential energy, which gets converted to kinetic energy, that is, energy of motion, when you release the object.

Guanine One of the nucleotide bases of which DNA is comprised.

Guide star (real or artificial) A star (if real) that is in the field of view of the telescope that is used to adjust the mirror so as to optimize the resolution. If the guide star is artificial, it is generated by a laser beam that excites atoms in the Earth's atmosphere, which then appear as a star, and which allow the resolution optimization.

Hayabusa A Japanese space mission that has landed on comet Itokawa, and has returned to Earth with samples from that comet.

Helium burning The second stage of stellar evolution, in which, primarily, three helium nuclei are converted into a carbon nucleus, and an additional helium nucleus can be added to make an oxygen nucleus.

Homochiral Having only one chirality. A homochiral medium is 100% enantiomeric.

Hubble's law The relationship $v = HR$ that relates the velocity of recession of distant objects to their distance.

Hubble constant The proportionality constant, H, in Hubble's Law, currently at 70.8 kilometers per second per megaparsec.

Hubble space telescope The telescope that was flown into space so as to circumvent the absorption of ultraviolet and infrared light by Earth's atmosphere, and also to avoid the resolution spoiling resulting from motion of the Earth's atmosphere.

Hydrodynamics Description of matter, through physics equations, as if it were a continuous fluid.

Hydrogen burning The first stage of stellar evolution, in which hydrogen induced nuclear reactions convert four protons into a helium nucleus.

Hydrothermal vent A vent in the ocean floor through which a variety of chemicals can be emitted.

Hyperbolic comet A comet that is made of stuff that is not from our Solar System. It would make only one pass through the Solar System, on an hyperbolic orbit, which is not closed on itself, so would originate outside the Solar System, pass through it, and then return to extrasolar space. Thus the material from which the hyperbolic comet was comprised would not necessarily have originated from the same region of space that the material in the Solar System did.

Hyperthermophiles A form of extremophile that can withstand very high temperatures, 80°C to as more than 120°C.

Infrared light Light that has a longer wavelength than visible (optical) light, that is, longer than 700 nm. The infrared region extends to 300 micrometers.

In phase This describes two oscillations, for example, two components of the electric field, that reach their respective maxima and minima simultaneously.

Intracluster medium This is the superheated gas near the center of a galaxy cluster. It contains mostly hydrogen and helium, and strongly emits x-rays.

Interstellar medium The material that exists in the space between the stars.

Ions or Ionized atoms Atoms or molecules that are not electrically neutral by virtue of having lost one or more of the electrons that would make them electrically neutral.

Isotope Isotopes of an element have nuclei with the same number of protons, but differing numbers of neutrons. Examples could be some of the isotopes of tin: ^{112}Sn, ^{114}Sn, ^{115}Sn, ^{116}Sn, ^{118}Sn, ^{120}Sn; or the stable isotopes of oxygen: ^{16}O, ^{17}O, and ^{18}O.

Itokawa The comet from which the Japanese space mission Hayabusa brought back samples to Earth.

Juan de Fuca ridge Location of a hydrothermal vent field in the northeast Pacific Ocean.

Junk DNA Sections of human DNA that do not fulfill any obvious physiological or replicative need.

Kelvin temperature A measurement of the temperature in which 0°K is absolute zero. Room temperature on this scale is about 300 K. Water freezes at 273.15 K, and it boils at 373.15 K.
Kepler space telescope A telescope that is primarily dedicated to finding extrasolar potentially habitable planets.
Kingdoms The classification of living things that includes archaea, bacteria, and four classes of eukaryotes.
Lepton conservation One of the conservation laws of physics that says that in any reaction or decay process the number of leptons must be conserved.
Leptons These are particles that don't interact via the strong interaction, so are not comprised of quarks. They include electrons, muons, tauons, neutrinos, and their antiparticles.
Life signals Detectable entities or events that would indicate the existence of living beings.
Light curve The temporal evolution of the light output from any star, for our purposes, a supernova.
Light year The distance light travels in 1 year. This is 9.47 trillion kilometers, or 5.9 trillion miles.
Linearly polarized light Light for which the electric field vector appears to oscillate in one direction (see figure in Chap. 4).
Lipids Organic compounds that are one of the components of living cells. They are greasy and not soluble in water. Their function is, among others, energy storage.
Local chirality The situation that exists when chirality is established in one place, but another chirality may exist at a different location. This would be the situation that would exist from chirality resulting from circularly polarized light.
Longitudinal polarization Situation that exists when the spin of a particle is parallel or antiparallel to its direction of motion. This is the case for photons and highly relativistic electrons.
Lookback time Astronomers' jargon for time measured backward from the present.
LUCA The Last Universal Common Ancestor.
Magnetic field That physical entity that affects moving charged particles and also, along with the electric field, describes the motions of light particles, or photons.
Magnetic moment A property of elementary particles, nuclei, and atoms that interacts with an external magnetic field to produce an

energy shift. Macroscopic quantities also have magnetic moments, for example, a current loop.

Magnetic monopole The magnetic equivalent of an isolated charge. If magnetic monopoles existed, magnetic field lines would not have to connect on themselves, but could, as with electric charges, radiate outward to infinity.

Magnetic substates The energy states of atoms or nuclei, or their combination, that result from the interaction of their magnetic moments with a magnetic field. The number of magnetic substates of an atom that has total angular momentum J will be 2J+1, and the projections of the vector J along the magnetic field direction will range from +J to −J in steps of 1 (in units of Planck's constant).

Magnitude The way in which astronomers describe the brightness of the objects they observe. An object that is one magnitude brighter than a second object is 2.512 times as bright as the second object. An object that is two magnitudes brighter than a second object is $(2.512)^2 = 6.310$ times as bright as the second object.

Mass energy This refers to a way of indicating masses by indicating them instead as the energy they contain according to Einstein's famous equation $E = mc^2$, where m is the mass of the object, and c is the speed of light.

Membrane (of a cell) This is the container that holds the components of the cell.

Metabolism The processes by which our cells sustain life. These allow cells to grow and reproduce.

Meteorite A meteoroid that penetrates the Earth's atmosphere and hits the ground.

Meteoroid A chunk of rock, and possibly ice, that exists as part of the interstellar medium.

MeV A unit of energy. One MeV = one million electron volts = a useful unit in nuclear physics, since that is a typical nuclear energy. One erg = 0.625×10^6 MeV.

Microbial mat A mat of living entities observed in the Octopus Spring in Yellowstone National Park.

Miller-Urey experiment The famous experiment, conducted in the early 1950s, in which an electric discharge in the presence of a few simple chemicals was found to produce amino acids.

Mirror isomers Molecules that are identical in every respect, for example, melting and boiling temperature, solubility, but which have opposite chirality.

Molecular clouds The giant clouds of gas, dust, and stars that occur in the Galaxy.

Mothro hydrothermal vent field A hydrothermal vent field in the northeast Pacific Ocean.

Muon An elementary particle. Muons can be either positively or negatively charged. They are about 260 times as massive as electrons or positrons. They have a spin angular momentum of $\hbar/2$.

Murchison meteorite The meteorite that landed in 1969 in Australia that contained a variety of amino acids, all of which were racemic, had a left-handed chirality, or were non chiral.

Mutation Change in the DNA coding in a cell that makes it different from what it previously was through, for example, inexact replication or cosmic rays.

Nebula The cloud of gas and dust that is produced when a massive star explodes.

Neon burning The stage of stellar evolution that is between carbon burning and oxygen burning.

Neutral-current weak interactions Those weak interactions that can transfer energy to a nucleus, but do not of themselves change the proton and neutron number of the nucleus.

Neutrino An elementary particle that has no charge, a very tiny mass, and interacts only through the weak interaction. Neutrinos come in three flavors: electron, mu, and tau. Antiparticles exist for each of those flavors. They have a spin angular momentum of $\hbar/2$.

Neutrino oscillations These are the conversions between neutrinos of different flavors, due to the fact that the different flavors of neutrinos can change back and forth, or in some cases, just convert from one to another.

Neutron One of the basic constituents of atomic nuclei, and of neutron stars. Neutrons are composed of three quarks, one up quark and two down quarks. They have a spin angular momentum of $\hbar/2$.

Neutron closed shells The filling of the quantum mechanical shells in a nucleus; these exist for nuclei that contain 2, 8, 20, 28, 50, 82, or 126 neutrons.

Neutron star A star that has collapsed essentially to the density of nuclei, and is comprised mostly of neutrons and a small fraction of protons. A neutron star is supported by the degeneracy pressure of its neutrons and protons.

Nuclear excited state A state of existence of a nucleus that exists at a higher energy, that is, is less tightly bound, than the ground state, or lowest energy state.

Nuclear reactions The interactions between two nuclei that result in different nuclei, either as a composite of the first two or as two or more new nuclei.

Nucleic acids See DNA and RNA.

Nucleobases These are the building blocks of RNA and DNA: adenine, cytosine, guanine (both DNA and RNA), thimine (DNA only), and uracil (RNA only).

Nucleosynthesis The creation of the elements via nuclear reactions. This mostly takes place in stars.

Octopus Spring A geothermal spring in Yellowstone National Park, located a few miles from Old Faithful geyser.

Olber's paradox The assertion that the night sky ought to be bright if the Universe is infinite and uniformly populated with stars.

Oligomers These are molecules that consist of several smaller molecules joined together. They are in contrast to polymers, which can have an unlimited number of smaller molecules.

Optical light Light having wavelengths that can see by the unaided eye, that is, with wavelengths from 400 to 700 nm.

Organic molecules Generally molecules that contain carbon. However, there are a few exceptions of molecules that contain carbon that are classed as inorganic molecules.

Oxygen burning The stage of stellar evolution that occurs between neon burning and silicon burning.

Panspermia hypothesis The hypothesis that life was created in outer space, or at least somewhere else, and then transported to Earth.

Paramedium bursaria Very unusual pond dwelling paramecium.

Parity Handedness. Parity is also a fundamental symmetry of physics that is violated by the weak interaction, but is conserved in most cases.

Parsec Astronomical distance unit = 3.26 light years.

Pauli principle The "law" of physics that requires that some types of particles in some systems, for example, electrons in atoms, or protons and neutrons in nuclei, have only one such particle in each allowed quantum mechanical state. This produces the well-defined sizes of atoms, of White dwarfs, and of neutron stars.

Peptides Relatively short protein-like chains of amino acids, but with different terminations than proteins have.

Period of oscillation The length of time for an oscillation to complete one full cycle.

pH Indicator of acidity or basicity of a medium. Zero is essentially purely acidic, seven is neutral, and fourteen is essentially purely basic. From the definition of pH, however, it is possible to have negative pH and pH greater than 14.

Photon A fundamental "particle." This is the particle of electromagnetic energy, and includes radio waves (very low energy), light (moderate energy), X-rays (higher energy), and gamma rays (extremely high energy).

Photosphere The periphery of a star.

Photosynthesis Production of complex organic materials, especially carbohydrates, from carbon dioxide, water, and minerals using sunlight as an energy source.

Pink filamentous Characteristics of a community of life found in Octopus Spring.

Pink streamers Description of a community of life found in Octopus Spring.

Pion A strongly interacting particle that is comprised of a quark-antiquark pair. There are three charge states: positive, neutral, and negative. They have a spin of zero.

Planck's constant This is the unit in which microscopic angular momenta are measured. Its value is 6.626×10^{-27} erg-seconds. It is also the unit that occurs in the famous Heisenberg Uncertainty Principle, which states that there is a fundamental limitation in the accuracy to which certain pairs of variables, for example, position and momentum, can be simultaneously measured.

Planetary nebula This is a glowing shell of ionized gas and plasma ejected during the asymptotic giant branch (helium burning) phase of certain types of stars.

Polarized light Light that has a specific orientation for its electric field vector (linearly polarized light) or an electric field vector that rotates in direction in time (circularly or elliptically polarized light).

Polycyclic aromatic hydrocarbons Complex molecules (generally more complex than amino acids) observed in outer space.

Polymers These are long molecules that consist of shorter segments called monomers.

Positron An elementary particle that is the antiparticle of the electron. It has the same mass as the electron, but has a positive charge. They have a spin angular momentum of $\hbar/2$.

pp-Chain This is the set of reactions that defines how hydrogen nuclei get fused into helium nuclei in not-too-massive stars.

Primary process This denotes a process of nucleosynthesis that creates a new set of seed nuclei each time it operates, and so produces the same results in stars no matter how many heavy nuclei they had previously. The r-process is primary, the s-process is not.

Progenitor star The star that exists prior to the supernova explosion.

Prokaryotes Not eukaryotes, that is, archaea or bacteria.

Protective mechanisms The means by which extremophiles shield themselves from extreme environments in which they live.

Protein Complex organic molecules on which we depend to sustain our lives. They are made from amino acids.

Protein world The view that the first molecules from which life eventually evolved must have been proteins, since making molecules as complex as RNA would have been too unlikely.

Protista Unicellular protozoans and multicellular algae.

Proton One of the basic constituents of atomic nuclei, and to a small extent, of neutron stars. Protons are comprised of three quarks: two up quarks and one down quark. They have a spin angular momentum of $\hbar/2$.

Quantum mechanics The theory that describes microscopic nature. It results in quantization of many entities that exist in steady state situations, such as energy and angular momentum, in atoms and nuclei.

Quantum state A characterization of all the properties of existence for a particle, including location, momentum, energy, angular momentum, spin, and whatever other variables are needed to completely specify the state.

Quark A fundamental particle. Quarks cannot be found by themselves, but either occur as three quarks together in particles like protons and neutrons, or as a quark-antiquark pair in mesons like the pion. Quarks have a spin angular momentum of $\hbar/2$.

Racemic Referring to an assembly of chiral molecules that has equal populations of the two possible chiral states.

Radiation pressure Pressure created by the outflow of photons from the center of a star.

Radicals These are atoms, molecules, or ions with unpaired electrons or an open electronic shell. They may be positively or negatively charged, or neutral.

Radiocarbon dating Use of the radioactive nucleus ^{14}C to determine ages of objects, e.g., wood and bone, that contain carbon. This technique only works for ages up to a maximum of less than 100,000 years.

Radio waves Electromagnetic radiation that has long wavelengths (greater than 1 mm) and correspondingly low energies.

Rankine One of several temperature scales. On this scale, absolute zero is 0°R, water freezes at 491.67°R, and it boils at 671.67°R.

Reaction probability The likelihood that a reaction will occur. See cross section.

Red giant A star that occurs in helium burning, when the core of the star generates more energy than it previously had, and the radiation pressure forces the periphery of the star to larger distances. This results in a larger surface area that, since the energy emitted by each unit area is reduced, appears redder than the star was in its hydrogen burning phase.

Red shift The shift that occurs in spectral lines from atoms or ions resulting from the source of those lines moving away from the observer.

Repair mechanisms The means by which cells repair components, especially DNA, that have been damaged by the extreme environments in which they live.

Replication Reproduction; this applies to living beings, cells, DNA, etc.

Ribosome The site within a cell at which protein assembly from the amino acid constituents occurs.

RNA Ribonucleic acid. One of a set of molecules that are involved in carrying on the functions of life, including assembling the proteins in cells. They include messenger RNA, or mRNA, and transfer RNA, or tRNA.

RNA world The view that the first molecules from which life evolved must have been RNA, since without RNA the proteins do not have their instructions for formation.

ROSETTA The European Space Agency mission that will send a lander to comet Churyumov Gerasimenko to analyze samples, among which will be, hopefully, amino acids.

r-process The process of nucleosynthesis that involves an extremely rapid sequence of neutron captures, synthesizes half the nuclides heavier then iron, and all the nuclides heavier than ^{209}Bi.
SETI Search for Extraterrestrial Intelligence. This usually refers to a search for radio signals that would characterize extraterrestrial intelligent life.
Shock wave This is a propagating disturbance in a solid, gas, or plasma. When a star collapses to slightly more than nuclear density, it first overshoots that density, then produces a bounce, thereby producing an outward going shock wave in the material that is external to it.
Silent supernovae These are core-collapse supernovae that ultimately collapse to a black hole, although they might first collapse to a neutron star. However, in either case, the black hole will swallow most of the electromagnetic radiation before it can escape the star. Neutrinos, however, will be emitted, and will escape before the black hole is formed.
Silicates Inorganic molecules that contain silicon.
Silicon burning The final phase of stellar evolution, which follows oxygen burning, and precedes collapse to a neutron star or black hole.
Smectite clay Clays are built of two types of sheets, each defined by its chemical structure. Smectite clays have a sheet of one type sandwiched between two sheets of the other type.
SNAAP model The model that describes how the amino acids achieved their chirality; SNAAP model stands for the Supernova Neutrino Amino Acid Processing model.
Solar system The system of planets and our Sun.
Spallation neutron source A facility for producing intense neutron beams, located in Oak Ridge, Tennessee, USA. It may also be used to produce intense neutrino beams.
Spectrograph An instrument that allows astronomers to resolve the different wavelengths of light emitted by a star. These can be used to identify the abundances of the elements in the photosphere of the star.
Spectrum The distribution of energies of the electromagnetic radiation over its possible range. This can also be the distribution over the range of wavelengths. Spectrum can also refer to a distribution of energies of particles. The plural of spectrum is spectra.

Spin An intrinsic degree of freedom of particles and nuclei that is a form of angular momentum.

Spitzer space telescope A space borne observatory of infrared radiation. It is sensitive to light of wavelengths which are absorbed by the Earth's atmosphere, hence the reason it is in space. It was named after Lyman Spitzer, Jr.

Spores Single walled or multiple celled reproductive bodies of organisms that are capable of surviving for long periods in hostile conditions.

s-process The process of nucleosynthesis that involves a slow (compared to typical beta-decay half-lives) sequence of neutron captures and beta-decays, synthesizes half the nuclides heavier then iron, and terminates at ^{209}Bi.

Standard candle A star that has some feature that allows astronomers to make an absolute determination of its distance. Examples are Cepheid variables, which have an oscillating luminosity that is related to its absolute luminosity, and Type Ia supernovae, all of which have approximately the same luminosity.

Standard solar model The model that describes the details of operation of the Sun. This includes all the complex hydrodynamics necessary in this description, as well as the nuclear reactions, photons and neutrino opacities, and other features necessary for a complete description.

Starburst superwinds These are winds that are generated from regions with an unusually high density of supernovae—the starburst regions. The superwinds are several times greater in mass, however, than could be generated by the supernovae alone.

Stardust A NASA mission to sample comet 81P/Wild2, and return to earth with some of its material.

Steady state universe The cosmological theory that the Universe had no beginning, but operated in a steady state. This theory is no longer viable.

Stellar evolution The different stages of nuclear burning that exists in massive stars, along with the concurrent hydrodynamic conditions that exist.

Stellar winds The winds by which some stars expel their outer layers into the interstellar medium. These are the result of radiation pressure from the photons being produced in the star.

Strong force/interaction One of the basic forces or interactions of physics.
Supernova These are the stellar explosions that produce so much light that they can outshine entire galaxies. There are Type Ia supernovae, which are powered by thermonuclear processes, and blow up the entire star. Type II supernovae are core-collapse supernovae, and are powered by gravity. They produce a neutron star or a black hole. Type Ib and Ic supernovae are also core collapse supernovae, but they have shed their outer hydrogen, and perhaps helium, envelopes before they explode.
Supernova Cosmology Project The program of measurement of Type Ia supernovae at a range of distances. The data from the SCP demonstrated the existence of dark energy.
Supernova remnants These are comprised of the gas that contains the newly synthesized elements that the supernova expels into the interstellar medium.
Tau A fundamental particle that can be either positively or negatively charged, and is more massive than the muon. Taus have a spin angular momentum of $\hbar/2$.
Template In this context, a molecular segment that provides a pattern for more molecules to follow.
Terrestrial Of the Earth.
Thermal energy The energy associated with the motion of the particles in the medium. This is heat energy.
Thermonuclear runaway The condition that exists in which the heat generated by nuclear processes in a star cannot be compensated by an expansion of the star, which would cool it. This is what causes a type Ia supernova to explode.
Thermophiles A type of extremophile that lives at fairly high temperature: 60–80°C.
Thermophilic cyanobacterium One of the life forms observed in Octopus Spring in Yellowstone National Park.
Threshold energy The energy that must be supplied to a reaction in order for it to proceed.
Thymine One of the nucleotide bases of which DNA is comprised.
Time dilation This is the result of moving at speeds close to the speed of light, in which the decay lifetimes of particles can become much longer, or space travelers can live much longer.

Type Ia supernova A stellar explosion in which a white dwarf exceeds its maximum mass due to accretion of matter from a companion. This results in a thermonuclear runaway.
Ultraviolet light Light having a wavelengths shorter than those in the optical; light with a wavelength less than 400 nm (but not as energetic as x-rays).
Universe The collection of all the stars and galaxies that we believe exist.
Uracil One of the nucleotide bases.
Vector A pictorial representation of a physical quantity. It has both length and direction, with the length representing the strength of the quantity being represented.
Visible light Light in the wavelength range from about 400 to 700 nm, which is visible to the naked eye.
Wavelength of oscillations The distance over which an oscillation completes one cycle.
Weak interaction One of the basic interactions of physics. The weak interaction mediates the process of nuclear beta-decay, as well as the interactions of neutrinos with nuclei and other particles.
White dwarf A final state of a medium mass star in which the size is maintained by electron degeneracy pressure. These are comprised either of carbon and oxygen, or of oxygen and magnesium.
White smokers Hydrothermal vents that emit liquid carbon dioxide.
Wild 2 The comet, technically 81P/WILD 2, that was visited by NASA mission Stardust, and found to contain amino acids.
Wilkinson Microwave Anisotropy Project WMAP measured the fluctuations in the 2.7 K cosmic microwave background radiation, and obtained the most accurate (as of 2010) data on the Hubble "constant," the baryonic and dark matter densities of the Universe, and the component of dark energy.
Wind blown bubbles These are bubbles that result from the winds from massive (greater than eight solar mass) stars.
Wolf-Rayet star These are very hot massive stars that have shed either one or two of their outer layers. If the former, they will be characterized by nitrogen emission lines. If the latter, they will exhibit carbon and oxygen emission lines.

X-rays Electromagnetic radiation that is more energetic than ultraviolet light, but less energetic than gamma rays.
Yamanote line The mass transit train line that encircles Tokyo.
Zircons Zirconium Silicate, $ZrSiO_4$. This is produced in magma. Zircons are crucial for determining ages of rocks because they initially contain uranium, but not lead, to which the uranium decays. This makes it possible to determine ages of the zircons.

Index